WINGS FOR VICTORY

BOOKS BY SPENCER DUNMORE

WINGS FOR VICTORY

The Remarkable Story of the British Commonwealth Air Training Plan in Canada

SPENCER DUNMORE

Canadian Cataloguing in Publication Data

Dunmore, Spencer, 1928-
Wings for victory: the remarkable story of the British Commonwealth Air Training
Plan in Canada

Includes bibliographical references and index.
ISBN 0-7710-2927-6
1. British Commonwealth Air Training Plan.   2. World War, 1939-1945 – Aerial
operations, Canadian.   3. Aeronautics, Military – Study and teaching –
Canada.   I. Title.

UG639.C2D8 1994   358.4'15'0971   C94-931650-4

Typesetting by M&S, Toronto

The publishers acknowledge the support of the Canada Council and the Ontario Arts Council for their publishing program.

The support of the Government of Ontario through the Ministry of Culture, Tourism and Recreation is acknowledged.

Printed and bound in Canada. The paper used in this book is acid-free.

McClelland & Stewart Inc.
The Canadian Publishers
481 University Avenue
Toronto, Ontario
M5G 2E9

1 2 3 4 5   98 97 96 95 94

Contents

Foreword

by Group Captain T. G. ("Hamish") Mahaddie,
DSO, DFC, AFC and Bar, CzMC, C.Eng., FRAeS, RAF (Ret)

I devoured the manuscript for this book the moment I received it from the author, in the main because I had little knowledge of the BCATP or of the sheer drama of the original scheme and the subsequent political haggle that nearly wrecked it before it got underway.

Happily, I was involved with the BCATP's "after product" as a customer at a Bomber Command OTU and, later still, when I was in the position to select special graduates at the dockside at Liverpool and Greenock. It was also helpful to be a wartime CFS graduate in following trained personnel arriving in the U.K.

Please-to-remember that as the Group Training Inspector of the Path Finders, or as Don Bennett's "horse thief," as I was better known, my main function was to select and train the best possible aircrew from a spectrum of talent. I found some worthwhile ferrets in the Canadian scheme who would let me know when outstanding potential Path Finders would be arriving from Canada having graduated from the BCATP.

I was fortunate in being in control of the Path Finder Training Unit at RAF Warboys. Thus it was possible for me to assess these ex-trainees and decide whether they should have more training as u/t Path Finders.

I believe I can justly claim, both as a "satisfied customer" and as an OTU instructor and founding member of the Bomber Command Path Finders, to have tried and tested the "product" of the

7

BCATP and found it of superb quality. I'm glad that the story of
this great scheme is finally being told in full.

*Distinguished airman, raconteur par excellence, T. G. "Hamish"
Mahaddie joined the RAF in 1928 as a Halton apprentice. His outstanding qualities were soon recognized, for he became one of the
very few ground staff accepted for pilot training during the 1930s.
When war came, Mahaddie found himself in the thick of operations,
flying the lumbering Whitley bomber. After a period of instructing,
he embarked on a second operational tour, this time on Stirlings. A
founder member of the famous Path Finders, he was appointed
Group Training Inspector, responsible for the supply of first-class
crews for the force. He retired from the RAF in 1958, becoming a
widely respected aviation consultant, specializing in film and television work. He is the author of* Hamish, *his highly entertaining
autobiography.*

Prologue

Shirt collars scratched like sandpaper. Boots, great leaden things, squeaked like obstinate truck springs. Side caps, still stiff from storage, kept closing up, clamlike, on freshly shorn skulls. But raise a hand to keep the pernicious thing in situ, and you got a blast from the corporal. *Stand goddam still in the ranks!* Attempting to explain the reason for the movement only made things worse. *Bloody quiet in the ranks!* Another lesson learned: on parade, you didn't debate, you obeyed.

It took far too long to form up. The corporal's features darkened, lips and nostrils a-twitch. He glowered at the untidy lines before him that wriggled like fidgety serpents. He glanced heavenward, as if seeking divine assistance, then informed the recruits that they were without question the sloppiest, stupidest bunch of deadbeats it had been his misfortune to encounter in twenty-four years of faithful service, and if even one of them managed to become a member of aircrew in the Royal Canadian Air Force it would be one of the miracles of our time and he personally would pass on the glad news to Rome.

Notwithstanding the corporal's assessment – one he imparted to every batch of inductees – the young recruits had to be rated as the cream of the country's crop, sound in wind, limb, and nerve. They hailed from every corner of the land and from more distant places like Gainesville, Texas, and Altoona, Pennsylvania. Soon they would be joined by trainees with even more unfamiliar accents, men who wrote letters home to towns and villages called

Kirby Overblow, Yorkshire, Paraparamaumu, New Zealand, and Oodnadatta, Australia.

Volunteers all, they had come to fight for democracy. And to fly. Although probably fewer than one in ten of them had ever handled the controls of an aircraft, most fancied their chances of becoming pilots. A few hoped to earn navigators' wings. Some hardy souls warmed to the notion of manning gun turrets or crouching over bombsights.

A lengthy selection process and training regimen awaited the men in their new uniforms. Many wouldn't make it. Scores would find to their indignation that the air force didn't consider them pilot material. They would have to settle for some other trade. The training process would injure some and kill some more. A handful would discover to their chagrin that, although they had dreamed of flying since childhood, they loathed and feared the reality of it.

Those who passed all the tests would eventually take their places on another parade ground. It would be an occasion of some importance. A senior officer or visitor of rank would preside. A line of aircraft might form a backdrop. A military band would probably be there, the musicians puffing gallantly as they competed with the brisk breeze that was sure to blow.

One by one the trainees would march forward to be congratulated and to receive the symbols of their accomplishment: aircrew badges, to be worn above the left breast pockets of their tunics, the coveted wings that testified to their newly won skills.

They would be counted among the best-trained airmen in the world, products of an immense scheme that had its beginnings in the late thirties and became Canada's principal contribution to victory in World War II. Known as the British Commonwealth Air Training Plan, it produced more than 130,000 aircrew.

This is the story of the BCATP and of the politicians who negotiated it into existence, of the officers and airmen of the RCAF and RAF, and of the many civilians who made it work day by day. Above all, it is the story of the young men who entered the scheme as clerks and farmers, students and salesmen, and graduated as pilots, navigators, air gunners, air bombers, and flight engineers. Some graduates stayed in Canada to pass on their brand new skills to other students; most went overseas to join an air force that grew in strength and efficiency until it became one of the principal reasons why the Axis powers went down to defeat in World War II.

CHAPTER ONE

Genesis of the Plan

de Havilland Tiger Moth

"If the Battle of Waterloo was won on the playing fields of Eton, the historian of the Second World War may, with some justification, record that the air battle of Europe was won on the fields of the BCATP."
   – Wing Commander Fred H. Hitchins[1]

August 1939. Europe was poised on the brink of war for the second time in a quarter of a century. There seemed to be a chilling inevitability about it, as if the entire continent were part of some insane Greek tragedy. In Southampton, England, twenty-five-year-old RAF Flying Officer Ken McDonald boarded a transatlantic liner. He wondered about the war that seemed imminent. Was the youth of his generation to be sacrificed just as their fathers had been twenty-five years before? Would this war be a repeat of the last one, with massed armies slaughtering each other by the tens of thousands in hopeless assaults across muddy, shell-torn fields?

McDonald thought not. Air power would decide this war, he believed. But he wondered if he would see any of it. Just as the balloon seemed ready to go up, he had been ordered to leave England and journey to Montreal, Quebec, Canada. *Canada,* of all places. The Royal Canadian Air Force (RCAF), he had been told, was embarking on a huge expansion of its training facilities. He had to go and help set up navigation training there. McDonald was simultaneously frustrated and intrigued. Although a capricious fate had whisked him away from England on the eve of war, he was getting a free trip to a land he had never before visited. He knew little more about Canada than what he had seen in Mountie movies and in newsreels of the Royal Tour. Forests, prairies, wheat fields, the Rockies, big American cars, snow, sunshine, and lots of pretty girls . . . and, yes, French-speaking Canadians – somewhere or other. McDonald had been the navigation officer at No. 11 FTS, Shawbury, England. When he had got his posting to Canada, Johnny McKid, a Canadian at Shawbury, wished him farewell with the words, "Mac, you're going to civilization."

McDonald crossed the Atlantic with four other RAF officers: Dick Waterhouse, Jackie Mellor, Desmond McGlinn, and Leslie Smallman. As their ship tied up in Montreal, an RCAF Norseman on pontoons landed close by. Out stepped two officers, Bill Proctor and Harvey Jasper. They had come to collect the new arrivals. McDonald and his colleagues bundled aboard the single-engined Norseman, taking their hand luggage – and managing to squeeze Waterhouse's black Labrador, Pluto, into the small cabin. A few minutes later they were on their way to their first stop in Canada, the air force base at Trenton, on the shores of Lake Ontario. Here they found what McDonald remembers as "a splendid Mess filled with cheerful people." Vastly impressed by the friendly, hospitable attitude of the Canadians he met that day, McDonald began to think that perhaps Johnny McKid had a point.

Canada on the eve of World War II was a nation emerging slowly, almost hesitantly, from the worst economic depression in its history. For ten grinding, maddening years, Canadians had endured the hardest of hard times. Would the economy ever recover? Would good times ever return? Many thought not, unless the nation's politics took an abrupt turn to the left, or right – although, perhaps surprisingly, the majority of Canadians still seemed to have faith in the tried and true political establishment. In 1939, Canadians were fiercely proud to be a part of the British Empire – at least, English-speaking Canadians were. But there existed in the nation a disquieting undertow of intolerance. The small town of Ste Agathe, Quebec, was a hotbed of anti-Semitism, according to the newspapers. On August 1, 1939, the Toronto *Globe and Mail* carried a story about two hundred posters appearing in the town announcing that Jews were not wanted and should "scram while the going is good." Catholics, said the story, claimed that they must remain "masters in their own communities."[2] Jews and Christians were barely on speaking terms in Ste Agathe during those days leading up to the outbreak of World War II.[3] "Restricted" was not an uncommon word in the Canada of late summer 1939. Ron Cassels, who became a navigator with Bomber Command, grew up in Warren, Manitoba. The community included only two Catholic families and no Jews; nevertheless, Cassels recalls, the local minister regularly warned his congregation about both groups.[4] Even quiescent rural Canada was not totally immune from the virulent plague of anti-Semitism called fascism that infected so much of Europe.

If you had money in those last few weeks of peace, there were bargains aplenty to be found. Electric refrigerators for $169.50. Men's pyjamas for $1.39 a pair. Milk ten cents a quart. Bread six cents a loaf. A two-piece wool suit from Eaton's for $20. You could get a room with bath at Toronto's Commodore Hotel for $4 a night, a three-bedroom apartment for $15 a month. Advertisements excitedly announced the 1940 Packards and Hudsons, due "soon in your showroom"; and at the CNE (Canadian National Exhibition) air show, scheduled for September 6, 7, and 8, the "revitalized RCAF" promised to show off its new Hurricane fighters and Oxford advanced trainers (described as "bombers" in the ads).[5]

War? Not likely, according to the press. Didn't the visiting diplomat Baron von Christe-Lomnitsky assure Canadians that Hitler was just a "passing phase?" Canadian-born newspaper magnate Lord Beaverbrook agreed, declaring flatly that there would be no war. British MP Beverly Baxter pointed to the might of the British navy and air force and of the French army. France's "unity of purpose," Baxter went on, was "beyond praise." The British Lord Chancellor, Lord Maugham (Somerset's brother), told everyone that the situation was "nothing to get frantic about." But, just in case things got out of hand, the French were said to be producing 750 first-line aircraft a month to add to the eight thousand or more already in use. A formidable force, *L'Armée de l'Air*. Didn't that erratic little moustachioed guy in Berlin realize just how powerful the Allies were? Didn't he understand what would happen if he pushed his luck too far? Was he really a mental case, as some claimed?

As August drew to a close, a note of uneasiness began to creep into the newspaper stories. The press published pictures of historic London buildings being sandbagged and accounts of millions of children being evacuated from the cities. "A great calm" had settled over London, reported the *Hamilton Spectator*. ". . . It shows that under the surface character of a race the heart is still strong, and the spirit unflinching."[6] From Ottawa came conflicting signals, as business-as-usual stories ran beside disquieting reminders of what might soon come to pass: "The Cabinet has dispersed for the summer, and the Prime Minister is enjoying his annual retreat at Kingsmere. If the newspapermen manage to intercept him on one of his rare visits to Parliament Hill and seek some pronouncement on Canada's policy in the event of a great war in which the Mother Country may be involved, he refuses to enlighten them."[7]

The man leading the nation in those parlous days was a sixty-four-year-old bachelor, William Lyon Mackenzie King. Earlier that month he had celebrated twenty years as leader of the Liberal Party. A chubby little man with a jolly smile, he looked like an aging choirboy, cheerful and content with his lot in life. In truth, he was a complex, eccentric person, a tormented product of the Victorian age who in his younger days had waged a consistently unsuccessful struggle with a vigorous sex drive. He seemed to regard it as a chronic ailment sent by the devil to plague him. Although he was rumoured to have had innumerable affairs, principally with married women, and had patronized countless prostitutes over the years, he gave only passing thought to marriage. According to some, it was because he had long ago despaired of ever meeting a woman to compare with his dead mother, Isabel, daughter of William Lyon Mackenzie. (Interestingly, Hitler had a similarly exalted opinion of his mother.) Fascinated by things mystical, enjoying regular discussions with his long-dead parents – and equally moribund pet terriers – King possessed an unshakeable conviction that he had been selected by the Almighty to protect Canada from its many enemies at home and abroad. William Lyon Mackenzie himself had told him so, during one particularly notable seance.

In spite of his undeniable peculiarities, King was a canny politician, who would lead the country for almost a quarter of a century, longer than any other Western politician in modern history. It is hard to pinpoint the reason for his success; he was a tedious speaker who seldom said anything in half a dozen plain and simple words when he could think of half a hundred ambiguous ones – and he usually could. His presence could never be described as impressive. He had an awkward, flat-footed gait.

Throughout the Royal Tour, he had strutted about behind the king and queen like a corpulent puppy, ready to deliver obsequious words of welcome on cue, basking in the reflected glory of the royal presence – even though he was forever scribbling angry notes in his diary about the British and Americans and how they treated Canada badly or ignored it altogether.

Although the political situation was a worry, King believed that everything would work out for the best, largely because of his own diplomatic triumph of two years before. In June 1937, he did something no other major Allied leader ever did. He met Hitler face to face. British prime minister Stanley Baldwin had objected strenuously when told of the proposed visit. So had King's closest adviser, O. D. Skelton, the undersecretary of state for external affairs. King went anyway. He assured Joan Patteson, his close friend (and, some said, surrogate mother), that the "forces" working for peace would triumph in the end. British aristocrat Violet Markham, another female friend of long standing, had warned King not to allow himself to be hypnotized by the German dictator. The warnings did no good. As King talked to Hitler he found himself attracted by the "appealing and affectionate look in his eyes."[8] What's more, he quickly realized – no doubt with some excitement – that the German was a mystic like himself.

King claimed that the talks with Hitler and the other Nazi leaders were most interesting and valuable. One wonders. King had arrived in Berlin with nothing of substance to discuss with Hitler and company – despite widespread belief that he was an unofficial emissary of the British government sent to give the Germans some straight-from-the-shoulder talk about the dangers of playing fast and loose with peace. In fact, according to King's own records, the visit to Germany was "private." He was scheduled to journey to London to attend the Imperial Conference and the coronation of King George VI; so why not drop in to see the gang in Berlin while he was in the neighbourhood, so to speak? The only remarkable thing about the meetings is that they ever took place. Presumably

Hitler felt that enough positive propaganda would be generated to make the whole tiresome business worthwhile.

As an exercise in international diplomacy, the Hitler-King meeting was a waste of everyone's time. To judge from King's notes, the result was less a conversation than a brace of monologues by two of the world's masters of rambling rhetoric. King told Hitler about Canada's place in the British Commonwealth of Nations, explaining that Canada would support Britain in the event of war. Such a move would be totally voluntary, however; the nation was under no obligation to aid the mother country. No doubt stifling a yawn, Hitler declared that, as far as Germany was concerned, there was no need for war. All the members of his government had been through the last one, and none wanted another. There were no matters, he went on, that could not be satisfactorily settled by negotiation. The greatest danger to world peace lay in the spread of communism. Impressed, King declared that the Nazi leader presented his case calmly, moderately, and logically.

King found much to admire in Germany. In spite of those nasty things people kept saying about Hitler, it couldn't be denied that the man had brought his country out of a dreadful mess. In chaos a few short years ago, Germany now bustled, dynamic and purposeful. The factories worked at capacity; the labour force toiled in apparent harmony. A network of super-highways would soon connect every corner of the country; in a few months, "People's Cars" (the forerunner of the VW Beetle) would be nipping along them in their tens of thousands. The Führer said so. But while the factories and their products were remarkable, the *people* of Germany seemed to have been reborn. After a visit to Hitler Youth and Labour Services camps, King couldn't help but compare the bronzed, confident Germans with those dismal, defeated souls on the dusty prairies, or in England, or the United States.

Following the meetings, King set off on the return trip to Canada, having accumulated a signed portrait of Hitler in a silver frame and a melange of misconceptions about the Nazi leader and his regime. Hitler had bamboozled King as few had ever bamboozled him before. He convinced the Canadian prime minister that he, Hitler, was at heart just a simpleminded peasant representing no danger to world peace. It was an amazing performance. King, the seasoned politician, the victor of a thousand political intrigues, swallowed it hook, line, and sinker. "The impression which King formed of Hitler was one not only absurd but calamitous," wrote historian Bruce Hutchison. "It distorted all King's thinking about the human tragedy now about to open."[9]

At home, King's visit to Hitler's Germany rated as something less than a diplomatic triumph. "Canadians must wonder what of benefit to this country their Premier imagines we can learn from the Nazis. . . . Mr. King's very presence, as a representative of a democratic nation, must give aid and comfort to those who have established and are maintaining their rule through ruthlessness and persecution," huffed the *Toronto Evening Telegram*.[10]

King ignored such criticism while he blithely overestimated the effect of both his presence and his words on the German leader. King convinced himself that he alone among the world's statesmen had succeeded in forging a genuine friendship with Hitler, a friendship on which the peace of the world might well depend. He and the Führer understood each other, he believed. Where others had failed, this unlikely pair – the fussy, peculiar Canadian prime minister and the totalitarian dictator – could solve any problem in an amicable, man-to-man way. As the international situation deteriorated, King sent a number of notes to the German dictator, reminding him of what they had said to one another back in 1937. The notes appeared to have no effect. One wonders if Hitler even remembered King. In any event Hitler's behaviour became progressively less diplomatic as the thirties slid helplessly to their catastrophic finale.

King and his government had been pondering the possibility of war for some time. If it came, what role would Canada play? She would, *of course,* support Britain, the Mother Country. That more or less went without saying. But what form would Canada's support take? And how to avoid the mistakes made last time, with all their perilous political implications? In the 1914–18 war, Canada and the other dominions had been little more than a handy pool of manpower for Britain. Canadians by the thousands disappeared into the bloody maw called the Western Front. Eventually the slaughter led to the Canadian conscription crisis of 1917, that hideous political wrangle that came frighteningly close to tearing the fifty-year-old nation apart along the lines of language. Essentially, English-speaking Canadians favoured conscription, French-speaking Canadians didn't. A generation later, the wounds still hadn't healed. If war came, King wanted Canadian support for Britain to be concentrated on the air force. The British had long been asking for flying training facilities to be set up in Canada. Such facilities would require large numbers of RCAF airmen as instructors, administrators, and ground crews. A relatively small contingent of Canadian airmen – all volunteers – would serve overseas, and casualties would never approach the blood-bath levels suffered by the Canadian Corps in Flanders. So, King reasoned, the country would not have to go through the agonies of another conscription crisis. At the time it must have seemed a brilliant solution to a politically pesky problem. As it turned out, events outmanoeuvred King and his meticulously wrought strategies. Midway through World War II, he would face another conscription crisis, one just as divisive as the first.

Discussions between the RCAF and Royal Air Force (RAF) had been going on for some years, ever since the British began their rearmament program in the early thirties. London saw Canada as a handy source of manpower and as a place to train thousands of airmen far from European battle areas. The concept had worked well in the Great War. The Royal Flying Corps (RFC) had set up an

elaborate recruiting and training organization in Canada. Advertisements appeared, appealing to "Clear-headed, keen young
men, 18 to 30 years of age – those possessing fair education and
sturdy physique" to be trained as pilots for the *Imperial* Royal
Flying Corps, as it was then known in Canada. Cadets received
$1.10 per day (interestingly, the same *per diem* rate as that paid to
AC2s – aircraftmen second class – in World War II). The RFC
liked "colonial" volunteers (often referred to at that period as
"black" or "coloured" troops). The British perceived them as all
admirably hardy outdoor types, perfectly suited to the rigours of
flying.

> The reason that flying officers from overseas are success-
> ful in flying at altitudes is largely because they have not
> "coddled" themselves but have been accustomed to lead-
> ing a life in the open, wearing a minimum of clothing.
> Thus, they have inured their bodies to withstand dis-
> comfort arising from the cold. This means that when
> exposed to the cold high altitudes there is not the same
> tendency for them to use up bodily fuel extravagantly in
> order to keep the body warm. . . . The man who coddles
> himself, who likes to live luxuriously, too warmly
> clothed, who shirks a cold dip in the mornings, is not the
> man who will stand the strain of exposure, or fly well on
> long-distance trips. [11]

So wrote Dr. Graeme Anderson, an air force flight surgeon in
World War I. Although Dr. Anderson's words conjure up stirring
images of vigorous, self-reliant "colonials" taking to the air as
naturally as to the skating rink, the fact is that the majority of Canadian volunteers grew up in urban areas such as Toronto, in
homes far more efficiently heated than most of those in Britain.

In World War I, Canadian pupils learned the mysteries of aviation at schools controlled and commanded by the British. It was a
hazardous business. In 1917, the RFC in Canada experienced one

fatality for every two hundred hours of training, a shockingly high rate that, fortunately for all concerned, would become far lower by war's end. Some three thousand Canadians received their training in Canada during World War I. Upon graduation, most of them went overseas, virtually losing their Canadian identity in the ranks of British units; if their subsequent deeds merited mention in the press, chances were they would be referred to as British airmen. In all, more than twenty thousand Canadians served in the British air forces.

Since the Armistice, Canadians had played only a minor role in the day-to-day activities of the RAF. The Air Ministry had reserved a paltry two permanent commissions per year for Canadian university or Royal Military College graduates. Canadians with high-school education could apply for short-service commissions in competition with British applicants. In the boisterous, prosperous days of the twenties, few Canadians had been much interested. The situation changed in the thirties. The economic slump virtually wiped out the small Canadian aviation industry, while the RCAF shrivelled under a series of draconian budget cuts instituted by hard-up governments. Opportunities for air-minded Canadian youth dwindled. In 1931, the air force granted an appointment to only one graduate. Not surprisingly, more and more young Canadians tried to join the RAF – although the numbers were still insignificant, a mere thirty-six Canadians obtaining short-service commissions in 1934–35. The Air Ministry liked to have a token sprinkling of Dominion airmen in the RAF at all times; the policy helped to maintain a high standard of personnel and involved the Commonwealth countries in an active way in the defence of the Empire.

In January 1933, Adolf Hitler became chancellor of Germany. International tensions quickly mounted. The Air Ministry began to worry about manpower for the RAF. Inevitably, official eyes looked to the dominions. The Air Ministry suggested a scheme whereby Canadian, South African, and New Zealand air force cadets could be granted short-service commissions, serve five years in the RAF, then return home for reserve service. (Such a scheme was already in operation in Australia.) The proposal reached Ottawa in November 1934. It took seven months for the Canadian government to approve it. The intention was to bring the new scheme into operation at the beginning of 1937, but world events caused it to be overtaken by other, more ambitious schemes. The first of these suggested that Canadian candidates might be examined and selected in Canada, with short-service commissions being granted before the individuals sailed for Britain. Although National Defence approved the scheme without delay, the politicians vacillated.

Meanwhile, early in 1936, another proposal reached Ottawa via the office of the recently appointed Canadian high commissioner in London, the passionately Anglophile Vincent Massey – who, it was said, could make even the English feel unspeakably colonial.[12] "It has occurred to Lord Swinton [the British minister for air]," wrote Massey, "that in addition to the Canadian Officers who are being admitted to the RAF under the arrangements which call for their training in England, an additional number might be provided with their preliminary training in Canada and taken on the strength of the RAF after having obtained there certain flying qualifications."[13] No doubt London thought that with three proposals on the table, the Canadian government would respond with alacrity. But the number of proposals served only to alarm Ottawa. What started out as a simple plan of mutual co-operation was beginning to look like a major military commitment, which would create heaven knows what political problems at home.

During this period, Air Commodore (later marshal of the Royal Air Force) Arthur W. Tedder became head of training for the RAF. One member of his staff was a Scottish airman, Group Captain Robert Leckie. Leckie knew Canada. Born in Glasgow in 1890, he had emigrated to Toronto in 1907, where he worked for a shipping company. On the outbreak of World War I, he returned to Britain and joined the Royal Naval Air Service (RNAS). A highly successful pilot, Leckie participated in the destruction of two Zeppelins. He also demonstrated administrative ability, commanding No. 1 Canadian Wing in Britain and later working on the Canadian Air Board, that short-lived organization set up to control all aviation, military and civil, in Canada (it disappeared with the creation of the RCAF in 1924), after which Leckie resumed his RAF career. In 1936, with the threat of renewed German ambitions beginning to cause concern in the Allied camp, Leckie wrote Tedder a memorandum entitled "Notes on the Proposal to Establish a Flying Training School in Canada." This was the genesis of the British Commonwealth Air Training Plan (BCATP). He pointed out:

A Flying Training School formed in Canada may be said to be practically immune from enemy action. (It is assumed that war with the United States is unthinkable.) The presence of the United States as a neutral guarantees the security of Canada. I cannot visualise any circumstances under which the United States would tolerate the intrusion of a European Power into Canada. . . . In the event of a European war depriving us, even temporarily, of our sea communications, an FTS in Canada would be able to maintain itself satisfactorily by obtaining its supplies of aircraft, engines, and war-like stores from American sources. (Unless this is precluded by recent American legislation on the subject of supplies during war to belligerents.) In any event, ocean-borne supplies on the North Atlantic are likely to be more easily maintained than in the Mediterranean or Southern hemisphere. . . . The advantages that would accrue from the operation of an FTS cradled in a Dominion like Canada, which has its own Royal Canadian Air Force, and where esprit de corps is high . . . are obvious and need not be laboured.... At the end of the war 1914–1918, 11,410 Canadians were commissioned or cadets under training for the RAF. Since the total number of RAF officers was approximately 30,000, and as these included Technical, Stores, and Medical, it will be seen that, in point of fact, a large portion of the pilots in the RAF were Canadians. The athletic outdoor life led by the young Canadian makes flying the natural progression from his ordinary sport.... There is in Canada an excellent source of supply for short service officers for the RAF of a type better, in my considered opinion, than we are recruiting today, and the presence of an FTS in their midst would crystallise interest in the RAF and certainly produce excellent applicants if these are required. . . . How the scheme would be received politically I am not in a position to say, but I am certain that it would be looked upon with great favour by the senior officers of the Royal Canadian Air Force, and the Department of the Ministry of Defence. Personally I am of the opinion that it is a most desirable scheme and infinitely to be preferred to forming a further FTS abroad. [14]

Tedder liked his assistant's proposal and discussed it with the Canadian minister of national defence, Ian Mackenzie, who was visiting London at the time. An intelligent, perceptive individual, Tedder seemed to be far more conscious of Canadian sensibilities than some of those who followed him. Tedder assured Mackenzie that such a training scheme could be set up without sparking a lot of political trouble in Canada. There were several choices. It could be run entirely as a British operation, with Canada simply supplying the site, or, if Ottawa preferred, it might involve considerable numbers of Canadian personnel and quantities of equipment; alternatively, the operation might be run jointly. [15]

When King heard about the proposal, his political antennae twitched in alarm. A week later the Cabinet rejected the plan, declaring that "it would be inadvisable to have Canadian territory used by the British Government for training school purposes for British airmen." The word went to London: "It is the intention of the Canadian Government to establish training schools of its own. The situation might give rise to competition between governments in the matter of fields, pilots, equipment and the like."[16]

King had to proceed with caution if he was to avoid a first-class political brouhaha at home. While most English-speaking Canadians may have been in favour of the British setting up training schools in Canada, the francophone population would undoubtedly have objected vociferously. King had no intention of becoming involved in any such fight, the only result of which would be to cost him votes.

But the British were not easily put off. With each passing day, the Germans were becoming more bellicose. The clouds of war began to gather over Europe. Preparations had to be made, and it was clear to every armchair strategist that the availability of training facilities and airmen might mean the difference between victory and defeat. Although the urgency of the matter could hardly be doubted, the ensuing discussions and exchanges of notes and other missives between London and Ottawa must have set international records for torpor. Lack of understanding was a major factor. The perspectives of the participants was another. On one hand were the politicians in London, many of whom still thought of Canada as a colony and expected any proposal emanating from the Mother Country to be acted upon more or less without question. On the other, the Canadian prime minister, who, for all his eccentricities, was determined to protect Canada's status as an independent nation. But it was a nation with two totally opposing views of military involvement. King, perpetually haunted by the fear that chills the heart of every democratically elected politician – that of offending some segment of the electorate, losing votes, and, horror of horrors, failing to win re-election – was in an unenviable position. Adding to the confusion, for much of the next three years neither side seemed quite sure what the other was talking about. Was it being suggested that Canadians *and* British airmen were to be trained in Canada or was it to be Canadians only? Or British only? No one seemed to know.

Early in 1937, the Canadian government announced approval of a plan proposed more than two years earlier to grant RAF short-service commissions to fifteen Canadians a year. Each would receive elementary flying training in Canada before departing for Britain. During Ottawa's interminable deliberations, the political situation in Europe had deteriorated. London was by now talking of larger numbers of student pilots; perhaps groups of up to twenty Canadian candidates "arriving at regular intervals throughout the year and commencing as soon as practicable."[17] Ottawa didn't respond for many months, and when word finally reached London it was to the effect that the Canadian government preferred not to increase the numbers to more than twenty-five per year "so as not to prejudice the position in Canada should it be necessary at a later date to secure this type of candidate for service in the Royal Canadian Air Force."[18]

Early in 1938, as Hitler reached out and annexed Austria to the Third Reich – without an official word of protest from London, Paris, Washington, or Ottawa – King at last agreed to increase the numbers of Canadians entering the RAF to 120 a year. Ever cautious, he added a codicil: the arrangement should not be regarded as a commitment.

In Britain the RAF was expanding, hastily re-equipping its squadrons with Blenheim and Battle, Wellington and Whitley bombers, and the first of the monoplane fighters, the Hurricanes. Spitfires would start rolling off the assembly lines a few months later. Enthusiastic aircrew volunteers swamped recruiting offices, far too many for the small training organization then in existence to handle. No wonder the RAF wanted the Canadians to train their 120 recruits up to the elementary level before sending them to Britain. Any proposal that would relieve the facilities at home had the support of the Air Ministry. Indeed, some officials advocated sending RAF recruits to Canada for elementary flying training; a vigorous training program in Canada would attract more Canadian recruits to the RAF, they claimed. Moreover, such an arrangement would involve the reluctant dominion in the reality of the crisis facing the Empire.

Some four hundred Canadians became pilots in the prewar RAF. One was an adventurous twenty-two-year-old, Pitt Clayton, who saw an advertisement in a Vancouver newspaper offering short-service commissions in the RAF. Eager to fly and to see something of the world, Clayton replied. Soon he received word to go to the RCAF seaplane base at Jericho Beach, Vancouver, for his initial medical. After that it was on to Ottawa for his final medical. Then at last he journeyed across the Atlantic, the tab for the entire exciting trip courtesy of the Air Ministry in London. In England, Clayton took elementary flying instruction on Tiger Moths at a flying club at Desford, Leicestershire. He was still a civilian – "the RAF wanted to make sure you could fly before they commissioned you," he recalls. Clayton was in the final stages of his pilot training when World War II broke out.

In May 1938, a mission led by industrialist J. G. Weir headed for North America, primarily to look into the possibility of purchasing aircraft for the RAF. Weir had another assignment: to discuss the long-standing question of establishing schools in Canada to train pilots for the RAF. Clearly the matter had become urgent. The Air Ministry said that Britain would pay for the school and the training, but would leave the operation of the school to the RCAF. It was a significant step forward and a reflection of the RAF's anxiety about international events. It was expected that most of the students would be Canadian, but British pilots could, it was suggested, fill any vacancies that might exist. A little progress had at last been made.

King discussed the matter with Sir Francis Floud, the British high commissioner in Ottawa. It was during these talks that King seemed to display a chronic inability – or unwillingness – to listen to precisely what Floud was saying. Records of the talks, including King's own notes, leave no doubt that the principal British concern was the training of Canadian pilots in Canada for service with the RAF. Nevertheless, when King summarized the meeting for his Cabinet colleagues, he talked about *British* pilots being trained in Canada. Moreover, he still seemed convinced that the British intended to build *and run* an air training organization in Canada, a course of action which would, he felt, "force an issue in Canada at once which would disclose a wide division of opinion ... by, first of all, creating disunion in Canada, and secondly, prejudicing in advance the position that might be taken at a later time."[19]

Paralysis once again overtook the talks. Fortunately, Floud was able to inform London that the main object of the mission, the purchasing of aircraft, had been successful. Although this was good news, the British were increasingly uneasy as the Germans began to turn their attention to the Sudetenland. They needed Canadian pilots – and Canadian participation in the huge air-training scheme which they were sure would soon be a necessity.

A month after their first meeting, Floud and King again discussed the idea of training British and Canadian pilots in Canada. And again King became confused, the apparent victim of a kind of mental tunnel vision. He managed to absorb the fact that the British had accepted that the RCAF would control any training facilities in Canada, but with Floud he still talked only of British pilots being trained. Although London had considered the possibility of sending British recruits to Canada for elementary flying training, the Air Ministry's principal interest was in recruiting and training Canadians in Canada. When the Conservative leader, R. B. Bennett, questioned in the House of Commons the government's position on air training, King could declare his government's willingness to co-operate in an Anglo-Canadian program "to give in our own establishments the opportunity to British pilots to come over here and train."[20] Did King have some

Machiavellian political purpose behind his apparent confusion? Or was he genuinely confused? The Air Ministry thought it would be Canadian pilots; King thought, or said he thought, it would be British pilots.

Remarkably, this rather important point had still not been cleared up when Group Captain J. M. Robb, the head of the RAF's Central Flying School, arrived in Ottawa. He and indeed all the British negotiators had no doubt that they were in Canada to discuss the training of up to 135 Canadian pilots a year for service in the RAF – a number the Air Ministry would have dearly loved to increase dramatically. King fretted over the new plan and its implications to Canadian sovereignty; and besides this, the plan's size troubled him. Training several hundred pilots a year for the RAF might have a detrimental impact on Canada's own defence plans, he claimed. It would mean that Canadians would be recruiting and training more men for the RAF than for the RCAF. King feared that voters might take it that the country was becoming entangled in a military commitment.

On July 19, George Croil, the senior air officer (and later chief of the air staff), discussed with Robb the feasibility of recruiting and training three hundred pilots a year up to the intermediate level, after which they would go to Britain for advanced training. The British would carry most of the financial burden. Early in August, Croil, Robb, and L. R. LaFlèche, the deputy minister of national defence, submitted the proposal to King. The PM objected at once, describing the plan as a recruiting scheme for the RAF and a means of obtaining training bases in Canada. His temper rose another degree or two when he read his evening paper: the Ottawa *Journal* carried a detailed account of Robb's mission. Angrily, King decided that the Canadian defence department was more guilty than the Air Ministry of promoting the recruiting of Canadians; it was all tied in with the sordid business of "imperialistic" programs and armaments contracts. And although Britain seemed prepared to foot most of the not-inconsiderable bill, there was concern that the scheme would overshadow Canada's own interests, taking the highly qualified men that Canada herself would need in the event of war. It couldn't be denied, however, that the plan, if implemented, would bring into existence a training organization of significant size, one that might serve Canada well if war came. In addition, the training plan would create a reserve of qualified pilots and maintenance staff – and would create work for what was left of the country's aircraft industry. The air staff proposed combining RCAF and RAF training, thus demonstrating that the RCAF was "an equal partner in the training scheme, and in no way impelled by the Royal Air Force." [21] Students could be divided almost equally between the RCAF and the RAF. Canada would supply the airfields and buildings; the British would, at least in the beginning, provide all the aircraft and their spare parts. Other costs would be split in proportion to the numbers of Canadian and British airmen involved.

But King still fretted about the impact of the scheme on Canadian voters, large numbers of whom were just as isolationist as Americans. While King knew that, in the event of war, Canada would come to the aid of the Mother Country, he did everything in his power to ensure that at no time was the impression created that Canada was in any way *obligated* to help. Whatever Canada did, the decision had to be hers – and, equally important, it had to be *seen* to be hers.

But other factors added to his concern. In the years leading up to World War II, King became increasingly dubious about the nation's ability to withstand the political stresses and strains of a major conflict. According to historian Bruce Hutchison, King feared that "in Quebec's present humor," Canada might "fly apart," [22] disintegrating like faulty masonry under the first bit of pressure.

On the last day of 1938, King formally rejected the latest British proposal. By now, although the Air Ministry still regarded the idea of training British pilots in Canada and Canadian pilots in Britain as prohibitively expensive, the international situation would permit no further delay. Four months later, in April 1939, after more rounds of discussion, everyone agreed: Canada would train up to fifty British pilots for the RAF; at the same time, seventy-five Canadians would be trained in Canada for the RCAF. Like most international arrangements, it wasn't as simple as it sounded. RCAF student pilots generally took a ten-month flying training course. Now, to integrate with the recently revised RAF syllabus, the single course became three: elementary, intermediate, and advanced, each of about sixteen weeks. The first stage would be handled by civilian flying clubs, a practice destined to continue when the BCATP took over. Early in 1939, eight clubs, in Vancouver, Calgary, Regina, Winnipeg, Hamilton, Toronto, Montreal, and Halifax, received word that they had been selected to train air force students. It was good news indeed, for, like most businesses, the clubs had struggled to survive during the thirties. Now they would have a guaranteed supply of students and a client who could reasonably be expected to pay the bills, in full and on time.

The first batch of British student pilots was due in Canada the following September. War broke out before they could set sail.

A seance on September 2, 1939, with Joan Patteson at Kingsmere was interrupted – much like a news flash in a television program – by a message from Mackenzie King's late father. A Pole had just shot Hitler dead, announced Mr. King, Sr., and all plans for war would be shelved. Then came a telephone call. Sadly, it wasn't confirmation of the departed Mr. King's message. News had just arrived from London: the British ambassador to Berlin would shortly deliver an ultimatum to Hitler, demanding the withdrawal of troops from Poland. King returned to the table, disappointed but still confident that all would be well. "His father's spirit still asserted that Hitler was dead. . . . The sitting ended with his mother declaring, 'War will be averted.'"[23]

It wasn't. The next day, Britain and France declared war on Germany. King listened to President Roosevelt's broadcast to the American people, then sadly scribbled in his diary: "I came away from the radio feeling an almost profound disgust. It was all words, words, words, America keeping out of this great issue which affects the destiny of mankind. And professing to do so in the name of peace when everything on which peace is based is threatened. I was really ashamed of the attitude of the United States. Their word, at this moment, might have helped to save millions of lives."[24]

A cable arrived from British prime minister Neville Chamberlain, stressing Britain's need for airmen and urging an expansion of Canadian training facilities without delay. Like most Europeans, Chamberlain expected the skies to be black with *Luftwaffe* aircraft in a matter of days or even hours after the declaration of war. The German propaganda machine had worked wonders in the years leading up to 1939, generating widespread fear of the *Luftwaffe* and its capabilities. (In fact, it might be argued that the extraordinary success of Joseph Goebbels' propaganda prior to World War II contributed significantly to the nation's eventual defeat in the air. Haunted by images of an immense, all-powerful *Luftwaffe,* the Allies set to work to build up their air forces to staggering size and effectiveness. The BCATP would be a major element in that immense buildup.)

For once, King wasted no time in responding. He assured Chamberlain that Canadian training facilities would be expanded, adding that "a number" of partially trained airmen would be sent to the RAF, but pointing out that Canada might require these airmen to be returned to the RCAF "if the Canadian Government should later decide upon the organization of distinctive Canadian air units for service overseas."[25]

On September 10, one week after Britain's declaration of war, Canada followed suit.

In London, the Air Ministry urged their RCAF counterparts to push on with plans for training airmen in Canada. Already, within days of the outbreak of war, the British were talking about an annual requirement of eight *thousand* pilots. If such numbers seemed awesome, even bigger things were in the planning stages. On September 13, at the Canadian high commissioner's office in London, Vincent Massey met his Australian counterpart, Stanley Bruce. According to Massey's memoirs, the purpose of the meeting was to discuss "the disparity in force and other gloomy features"[26] of the balance of air power in Europe. After the meeting, the two men conveyed their uneasiness to Anthony Eden, the dominions secretary in Chamberlain's government. It was at that point, Massey later declared, that he had the idea that "Canada might be able to make a decisive contribution to the common war effort by training Commonwealth airmen." He also said that Bruce "enthusiastically agreed."[27] Thus, there can be little doubt that Massey considered himself the architect of the British Commonwealth Air Training Plan (he was almost certainly unaware of Leckie's 1936 memorandum). But Bruce had other recollections. "I was convinced that the conception of the plan was mine,"[28] the Australian told his biographer. Bruce claimed that the idea had occurred to him before the outbreak of war, but no documents have been found to back him up. According to Dominions Office records, the suggestion was made jointly by the Canadian and Australian representatives at a meeting with Eden on September 16. In any event, the whole business infuriated King. In his view, Massey had far exceeded his authority by entering into discussions on the proposed Commonwealth air training scheme without telling Ottawa. King described Massey's conduct as "outrageous."[29] One might say that King had only himself to blame. He had placed Massey at the very centre of world events and had given him virtually nothing to do. (At about this time, Massey noted in his diary that he had not heard from King since July.) An ambitious, energetic man like Massey could not be expected to sit and twiddle his thumbs in the midst of the most tumultuous events of the twentieth century.

In spite of these internal stresses and strains, King was generally positive when he heard about the proposed plan to train Commonwealth airmen in Canada. The basic idea dovetailed with his desire to contribute to the war while avoiding enormous casualties and disgruntled voters. It simultaneously irritated and pleased him that the British, so woefully unprepared for war, now expected Canada to come to their rescue. Canada, he informed London, would be happy to host a conference on the subject of a Commonwealth-wide air training plan. The British sent a draft statement concerning the proposed plan and asked Canada to approve it. Again King responded with uncharacteristic celerity, objecting only that the draft implied that the idea for the plan had emanated in Canada. King insisted on a clear statement that this was a British initiative, and, he declared, the statement "should also emphasize the fact that this activity constitutes . . . the most essential and effective form of military cooperation open to Canada."[30] Lastly, he rejected the suggestion that the details of the plan were all but settled. Besides, what would it all cost? No one in Ottawa seemed to have any idea what the country was getting itself into?

After his pleasant introduction to the RCAF at Trenton, Ontario, Ken McDonald of the RAF found himself at Camp Borden, north of Toronto, a grass airfield with a collection of World War I-vintage hangars. On the day Canada declared war on Germany, Wing Commander (later Air Vice-Marshal) Leigh Stevenson informed his officers – "pretty much all in one sentence," McDonald recalls – that Canada was at war, that it was his officers' job to lead, and that if they saw anything that needed doing they shouldn't wait to be told but should get on with it. The first few weeks of the war were hectic; everyone worked on a twenty-four-hours-on, twenty-four-hours-off basis in order to get the most flying time out of the available aircraft and instructors. Most of McDonald's "off" days were spent in Toronto, with the Royal York Hotel as the base of operations. He got along well with the Canadians he met, including the CPR engineers at Union Station, who generously provided him and Jackie Mellor, one of the RAF officers who had accompanied him across the Atlantic, with several highly enjoyable minutes of dual instruction on their gigantic locomotives. Back at Borden, McDonald listened in awe to the tales of the bush pilots then being trained as instructors. "We RAF pilots had much to learn from them," he admits. For their part, the Canadians were delighted to have the RAF officers to assist them. Joffre Woolfenden, a prewar RCAF pilot, considers that they made a "great contribution. . . . They had experience; we didn't," he remarks.

The declaration of war had a seismic impact on millions of lives. The familiar world had vanished. No one knew what the new one had in store. On the day of Chamberlain's lugubrious announcement to the British people, a twenty-year-old officer cadet named Chester Hull arrived at the Royal Military College, Kingston, Ontario, to start his third year of studies. But now that the war had started, he found himself changing his mind. He decided to see about joining the air force. At the same time, two hundred miles away in Hamilton, Ontario, a young man named Dave Goldberg clambered aboard a streetcar and heard someone say that Britain and Germany were at war. Goldberg, who had just graduated from Boston University, began to wonder: army or air force? Which was it to be? He didn't wonder long. He chose the air force. Nearby, at the Hamilton Aero Club, the personnel of 119 Squadron lined up on parade. An auxiliary squadron originally authorized as 19 (Bomber) Squadron, it had been formed in 1935 with a complement of such dubious warbirds as the de Havilland Moth and the Fleet Fawn. The following year, eighteen-year-old Alan Ramsay had joined the unit as an "aircraft hand." For the next three years, Ramsay trained a couple of nights every week and at weekends, with three weeks at Camp Borden every year. By the outbreak of war, Ramsay had a good deal of experience of aero-engines, airframes, and armaments; he had risen to the rank of corporal and was in charge of the squadron armament section. Expecting to be shipped to Europe, the members of the squadron were surprised to find themselves posted to the West Coast to join 6 Squadron, a bomber reconnaissance unit engaged in coastal patrols with outdated Blackburn Shark biplanes and porcine Northrop Delta monoplanes. In London, Ontario, Paul Laskey, a moulder working for a metal casting company, lost no time in going to the recruiting office. He had been earning nine dollars a week, on which he supported his wife and one child, spending ten dollars a month for the rental of a small house. He was surprised to learn that, if accepted for the RCAF, he could look forward to a modest increase in his earnings.

Across the Atlantic, in Leeds, England, Tet Walston listened to Chamberlain's speech about the war while wondering how all this would affect his final interview for a short-service commission in the RAF. It had been scheduled for Tuesday, September 5, 1939. Since childhood, Walston had cherished ambitions of becoming a pilot. Although he was employed as a mechanical engineering apprentice, he cared only about flying. Now that the country was at war, he told himself, the whole process of joining the air force would probably be speeded up. Later that same day, a telegram arrived, informing him that the interview planned for September 5 had been cancelled: "All further recruitment by usual channels." Walston went to the recruiting centre. There, a man from the Ministry of Labour confronted him. This worthy's job was to ensure that no "service exempt" men attempted to join up. To his dismay, Walston found that his engineering apprenticeship barred him from enlisting. In vain he protested that he had almost completed the process of becoming a pilot with the RAF. The man from the Ministry of Labour merely shrugged. According to the rules of the game, Walston qualified as exempt, and exempt he was going to be, like it or not. Walston went home convinced that the world had gone mad. In peacetime, he had been accepted, more or less, for the air force. In wartime, he was being rejected!

At Scampton, Lincolnshire, the aircrews of 83 Squadron tested their Hampden bombers. The squadron's state of readiness could only be described as lamentable. One of its pilots, a young man named Guy Gibson, wrote: ". . . another man got up and told us how to take off with a bomb load on. None of us had ever done it before and we did not even know whether our Hampdens would unstick with 2,000 pounds of bombs."[31]

As that eventful September drew to a close, Mackenzie King cabled Chamberlain, advising him of Canada's agreement in principle to the proposed air training plan. At long last the die was cast; Canada was to become an immense air training centre. Chamberlain had talked about eventually training no less than *fifty thousand* Commonwealth aircrew a year, nearly half of that number to be raised in Britain, the remainder to come from the Commonwealth and elsewhere. The scheme would undoubtedly have a profound influence on the course of the war, Chamberlain added, concluding hopefully: "The knowledge that a vast potential was being built up in the Dominions where no German air activity could interfere with expansion might well have a psychological effect on the Germans equal to that produced by the intervention of the United States in the last war."[32]

Now Canada had a hugely important role to play in the war. And, of great significance to King, it was a highly technical, modern role. The world would see a new Canada. But could the nation afford it? King was as cautious with the country's funds as with his own. Preliminary estimates by King's staff suggested that the first year of the war might cost Canada nearly half a billion dollars – and Canada had already offered Britain unlimited credit for the purchase of food and war *matériel*. Moreover, the balance of payment problem with the United States could only get worse as air training equipment was purchased in the huge quantities that would be necessary. Where would it all end?

October 1939. The first month of the war had ended. Poland discovered to its cost that gallant cavalry with lances was no match for tanks. It capitulated. A U-boat sank the British liner *Athenia*. (One of the passengers, a young American named Jim Goodson, swore that he would join the RAF if he made it to shore. He did, and he kept his promise to himself to join the air force, becoming a leading fighter ace.)

Mackenzie King awoke on the morning of October 23, noting in his diary that he had had a "vision" (King never had mere dreams, always visions). He had seen an airplane shooting forth "what looked like a series of banners tied together with strings. One large piece seemed to fly from the skies down in the direction of where I was. . . . My brother Max was nearby. I remember running out into the field to get it before he could reach it." King took it all to signify "the power of the aeroplane in determining ultimate victory."[33]

En route to Canada, the British representatives to the forthcoming air training talks, led by industrialist Lord Riverdale, discussed their proposals. They envisaged a plan which would initially produce some twenty thousand pilots and thirty thousand other aircrew a year, with each dominion providing elementary flying training facilities and Canada bringing all the students up to advanced flying level. The plan would require twelve elementary flying training schools (EFTSs) in Canada, twenty-five advanced flying schools (SFTSs), fifteen air observer schools

(AOSs), fifteen bombing and gunnery schools (B&GSs), three air navigation schools, and one wireless training school. About five thousand aircraft would be required, with some fifty thousand training and maintenance personnel. At this point, Riverdale anticipated that the RAF would run the plan; privately he believed that it was beyond the RCAF's capabilities to "organise and control a training scheme of this magnitude."[34] He also hoped to convince the dominions to foot most of the bill. These hopes would soon be dashed. Already the anticipated costs of the plan were sparking some acrimony between London and Ottawa. Confronted by an Everest of war bills, the British wanted Canada to supply wheat at the lowest possible price. But the Canadians wanted higher prices, and for precisely the same reason: the Canadian war effort had to be paid for.

Although Riverdale knew Canada and Canadians well, he had a tendency to think in colonial terms. No sooner had he arrived in Ottawa than he antagonized King by referring to "your scheme" and talking to the press as if the essentials had been worked out and his job was simply to explain to the Canadians and other dominion representatives what he expected of them. Tempers began to simmer even before the conference got off to its uneasy start. Although King managed to comport himself in a diplomatic manner, his private comments to his diary were less amiable, talking about the "railroading, taking-for-granted style" which Riverdale adopted, adding that it was "amazing how these people have come out from the Old Country and seem to think that all they have to do is tell us what is to be done. No wonder they get the backs of people up on this side."[35]

Riverdale and Harold Balfour, the British parliamentary undersecretary for air, met with the Canadian Cabinet on October 31. The BCATP nearly came to a premature end there and then. Riverdale airily proposed that Canada bear some 40 per cent of the cost of the plan in addition to providing most of the trainees. For its part, Riverdale promised, Britain would supply aircraft, engines, and parts worth approximately $140 million. Riverdale described this as a "free contribution." To the Canadians, the proposal had a patronizing ring. Angrily, King said the British regarded the whole exercise as "a recruiting scheme for the British Air Force rather than any genuine attempt for any cooperation."[36] Then came the sticker shock. Based on an estimated three-year duration, the plan would cost about $900,000. Aghast, King said the costs could not be allocated as Riverdale had suggested; the whole thing was, after all, "a scheme suggested by the British government for which the British must be mainly responsible."[37] Backing up his boss, Colonel J. L. Ralston, the Canadian minister of finance, said Britain's contribution was far too small. Did the British want Canada to be "financially bled to death"?[38]

Early in November, the Australian and New Zealand delegations arrived. If Riverdale and King expected their presence to make things easier, they were rapidly disillusioned. Within a few days, the representatives from the antipodes began complaining about receiving second-class treatment and being forced to wait in anterooms until the British and Canadians were ready to receive them. What's more, they were now nursing serious reservations about the whole plan. For one thing, Canada wanted payment in a form that could be readily converted into U.S. dollars; Australian and New Zealand currency reserves were strictly limited. The Australians and New Zealanders declared that their participation in the plan depended upon three conditions: first, that Britain underwrite their monthly payments to Canada; second, that the contribution of aircrew candidates be recalculated on the basis of populations – in other words, 57 per cent from Canada, 35 per cent from Australia, and 8 per cent from New Zealand; third, that Australia be allowed to train most of its aircrew recruits at home. The Australians and New Zealanders seemed to be in no mood for protracted negotiations on these points; they threatened to head home by the next available ship if their proposals were rejected.

The clerks hurried back to their adding machines and cranked out new sets of numbers. The politicians gulped, groaned, pondered, but finally nodded. Now the proposed scheme was shrinking a little. Sixteen advanced flying training schools instead of the twenty-five originally envisioned; ten instead of fifteen air observer schools and bombing and gunnery schools, two instead of the original three air navigation schools.

Moreover, it was calculated that the scheme would require only 3,540 aircraft instead of the 5,000 first considered essential: 702 Canadian-built Tiger Moths and Fleet Finches for elementary instruction, 720 North American Harvards for advanced single-engine instruction, 1,368 Avro Ansons for twin-engine pilot training and for navigation instruction, and 750 Fairey Battles for gunnery instruction. Britain promised to provide the Ansons and Battles, more than 500 of the Harvards along with 133 replacement engines, as well as half of the engines for the Tiger Moths. The rest of the Harvards were to be provided by Canada, Australia, and New Zealand. Canada would also pay for the Tiger Moths, half of their engines, plus the Finch airframes and engines.

The estimated cost of the air training plan in its final form amounted to rather more than $600 million. This budget allowed for funds to last until March 31, 1943, by which time, it was hoped, the war would be over. Britain would contribute aircraft and parts to the tune of $185 million, leaving a balance of more than $420 million. It was proposed that Canada contribute in excess of $66 million in the form of initial and elementary training for Canadian and British pupils. The remainder, some $350 million, would be divided among the three dominions: $287,179,331 for Canada, $40,170,918 for Australia, and $28,774,913 for New Zealand, based upon the numbers of trainees the nations were expected to contribute.[39]

King insisted that the RCAF run the training plan; the other Commonwealth nations would have a voice in the day-to-day operations of the plan through a monthly supervisory board meeting. In addition, each country had an air liaison officer in Ottawa, but the overall control had to be in Canadian hands. While in Canada, trainees from the other countries would be "attached" to the RCAF, an arrangement greatly simplified by the basic similarities between the air forces of the nations involved. Whatever their doubts about the RCAF's ability to run the training plan, Riverdale and his colleagues raised no objections.

They were finally close to complete agreement. The British were understandably keen to wrap up the last details. But King insisted on two provisions: the satisfactory completion of financial talks then going on in London between British officials and T. A. Crerar, the minister of mines and resources; and a statement from the British asserting that the BCATP was the most important of all Canadian military commitments. King needed such a statement for home consumption, for the inevitable battles in Parliament and as something to wave in the face of angry voters who might complain that Canada's ground forces should be larger and more powerful. It took time to arrive at the right wording. Obligingly, Kingsley Wood, the British secretary of state for air, declared in the House of Commons that the BCATP "may in the opinion of the United Kingdom Government prove to be a contribution of the most essential and decisive character."[40] King wanted something stronger, something including the statement: "The air training plan should take priority over all other Canadian commitments not already entered into."[41] Chamberlain provided a paragraph that said just that; but he added a few lines to the effect that large Canadian land forces would also be welcome.

By December, the essential features of the plan were in place. The participants agreed that the plan should produce 1,464 trained aircrew per month. The RCAF would be in charge, with assistance from the RAF. However, little thought had been given to how all these trained airmen would be employed when they graduated. On this point, as on so many others, the lines of communication between the interested parties were convoluted at best, totally ineffective at worst. The British had assumed all along that the graduates of the plan would be at their disposal, to be used as the Air Ministry saw fit. But the dominions had other ideas; they liked the idea of having their own squadrons clearly identified as such and manned by their own airmen. Canada wanted even larger units to be identified as Canadian, an ambition realized with the creation in 1943 of 6 (RCAF) Group, which by the end of the war represented about one-fifth of the strength of RAF Bomber Command. Indeed, the Air Ministry had already agreed in principle to the eventual creation of such units, although Riverdale probably wasn't aware of it.

It's hard to escape the feeling that the British simply wanted to get the air training plan running; the lofty ambitions of Commonwealth colleagues were something to worry about later. Neither Riverdale nor any of his advisers appreciated how strongly the dominions felt about the creation of national squadrons. They soon found out. Mackenzie King was adamant that the British agree to the Canadian graduates of the plan being organized in Canadian squadrons; it was "essential to Canadian participation in the Scheme." [42] In the event, however, only a fraction of Canadian BCATP graduates ever served in Canadian squadrons.

The training plan agreement between Britain, Canada, Australia, and New Zealand – now known as the British Commonwealth Air Training Plan, by far the largest part of the overall Empire Air Training Plan – became a reality in the early hours of December 17, King's sixty-fifth birthday. The communication problems persisted until the last moments. Still embroiled in the complexities of establishing the "Canadian" squadrons overseas, concerned that keeping most of its ground staff in Canada to look after BCATP aircraft might result in poor morale and perhaps problems in recruiting, the RCAF suggested that some maintenance staff be posted overseas. Why not bring RAF staff to Canada to replace them? King was outraged. A totally unacceptable idea, he declared, because it would inevitably "result in public criticism that Canadians were being substituted for UK personnel in zones of danger." [43] Moreover, he asserted, the Canadian government expected Britain to maintain the air force overseas, and that included the RCAF units with RAF ground crews.

Not for the first time, the British shook their heads in bemusement. For the past six weeks, each side had been confidently assuming that the other would pay for the RCAF's ground crews overseas. The situation was made even more tense when the Australians issued a statement announcing the creation of the training plan. What infuriated King was that the statement was issued through the Air Ministry in London – and it seems likely that London encouraged the whole thing to hurry the decision process along. King cabled London: "I cannot begin to express my amazement that without consultation with the Government of Canada, the United Kingdom Air Ministry should have concurred in the issue of the statement by Australia before agreement had been reached between all parties. I need hardly add that the publication has caused great embarrassment to our Government in relation to other Commonwealth Governments as well as the press and people of our own country." [44]

Chamberlain apologized, explaining that it was all a "regrettable misunderstanding." [45] The phrase might have been applied to the entire tangle of negotiations that had been inching painfully along since the mid-thirties. The BCATP, a momentous achievement by any standards, came into being in spite of rather than because of the politicians. King was undoubtedly relieved that the complex negotiations had been concluded successfully, and he seemed equally delighted at how well he had handled the overbearing Riverdale. The Englishman, King wrote, "saw that he had gone a little too far in his attempt at railroading" [46] – which King compared to Hitler's pressure tactics. King noted that Riverdale had been surprised at the efficiency of the Canadian bureaucracy, being "amazed at what had been accomplished in two months."[47] Intensely pleased with himself – a not untypical condition – King announced the creation of the British Commonwealth Air Training Plan to the Canadian people, telling them that it was a "co-operative undertaking of 'great magnitude'" that would "establish Canada as one of the greatest air training centres of the world."[48]

In general, Canadians agreed. The *Globe and Mail* commented: "It is a scheme around which Canadians, eager to turn their vigor to war channels, can rally with the greatest of satisfaction since it will require six months to train a pilot, nearly as long for an observer, and a slightly shorter period for air gunners and wireless operators. This is Canada's war, a war to preserve her institutions and privileges, and we cannot complain about essential, well-ordered expenses. The care with which the air training scheme has been worked out in a comparatively short time implies the expenses having had the fullest consideration. Substantial permanent benefits will arise from the training and the establishment of a great air centre in this country, but we have the word of the British authorities that by this means the Dominion can make its greatest contribution to the Allied cause, and this is what we wish to do."[49]

Soon the vast air training plan would be producing thousands of airmen. While Canadians discussed it and pondered its implications, Air Ministry officials in London worried about the Canadians' ability to run it. Did they have the personnel? The know-how? Could they carve out the airfields in time? Build the buildings? Find the instructors and ground staff?

Good questions. As yet no one had the answers.

CHAPTER TWO

The Grandiose Enterprise

Fairchild Cornell

"The most grandiose single enterprise upon which Canada had ever embarked – the British Commonwealth Air Training Plan"
   – Charles Gavan Power, minister of national defence for air during World War II [1]

At the outbreak of war, eager Canadians had lined up in their thousands to enlist: idealists burning to put their lives on the line for democracy, adventurers looking for a change from the tedium of everyday life, unemployed men seeking steady work . . . and those of thickening girth and thinning hair who had been through it all last time and yearned to serve again.

The newspapers had sent photographers to take pictures of the lineups and written about how the old spirit – the spirit of Vimy Ridge – still motivated Canadians. Canada's youth would never let the Empire down. But the pace of recruiting slowed after the first rush. The upswing in the economy was doubtless part of the reason; the war had delivered a gratifying shot in the arm to business of all types. Employment was up. "Help Wanted" signs began to appear on plant gates, reminders of that almost-forgotten time of prosperity before October 1929. And no doubt the situation at the front had a hand in it. For this was the time of the "Phoney War," that strange period in the last few months of 1939 and early 1940 when the conflict seemed to be in a state of suspension, with the armies of France, Germany, and Britain doing little but glaring at each other across miles of disputed territory. In bars and barber shops, citizens told each other that to all intents and purposes the war was over. Negotiations were going on behind the lines at that very moment, and the whole thing would soon be settled at the conference table. It made sense. And if indeed the politicians were sorting things out between themselves, there wasn't much point in signing up, was there?

But in mid-December 1939 came the prime minister's announcement of the creation of this huge British Commonwealth Air Training Plan. It seemed that the government expected a long war after all. Canada was about to become the world's biggest air training centre, with the RCAF expanding to unheard-of proportions and bases being built clear across the country. For thousands of young Canadians, this news was encouragement enough. Back to the recruiting offices they went, to wait in the winter cold, stamping their feet on the sidewalks, chatting with their neighbours in the uneasy yet elaborately nonchalant way of men on the brink of the unknown. This was unquestionably to be the greatest adventure of their young lives.

A few had tried to join the air force before the war. They had found to their dismay that only paragons possessing brilliant college degrees and the physique of a Johnny Weismuller need trouble to apply – and only a fraction of *those* stood a glimmer of a chance of getting in. Now, it was said, many of the peacetime restrictions would soon be eased, though the authorities seemed to be in no hurry. Fervent patriotism undoubtedly motivated many volunteers, but what made most of them hurry to the recruiting office was the chance of *flying*, of becoming that most exalted of all beings, an airman. Of course there were dangers. Everyone knew that. It was part of the appeal, the spice that made it irresistible. And it would be the other guys who would get into trouble, not you. You were sure of that. Or pretty sure.

That redoubtable ace of the Great War, Billy Bishop, was unquestionably the air force's most successful recruiter. He was a stocky, bemedalled figure, with a jolly smile and rows of medal ribbons, including the Victoria Cross. He kept popping up in newsreels and in magazines, still a national hero, a name to be reckoned with. It was a time when brave men and women were admired wholeheartedly, unreservedly. Their acts of courage inspired an entire generation. Countless young Canadians volunteered for the air force, nursing secret longings to be the Billy Bishop of this war, shooting down enemy aircraft by the score, to be lionized by society and decorated by the king.

At the same time, it is undoubtedly true that a major attraction of becoming an airman was that it was a good way of staying out of the trenches. Arthur Bishop of New Minas, Nova Scotia, (no relation to Billy) says of his decision to join the air force: "One uncle was killed by a shell blowing him to bits at Passchendaele and another was gassed in the trenches at Ypres, which provided enough family stories to discourage me from entering the army." Ron Cassels, who had been warned about Jews and Catholics before the war, had three uncles who had served in the army in World War I: "I had been told stories about the trench warfare, the mud and water, the lice, the gas attacks and the hardtack which substituted for food. If that was army life it wasn't very appealing."[2] The prospect of hand-to-hand combat appalled Marcel Beauchamp of Montreal: "I simply could not see myself shooting or bayoneting another human being."

Some recruits found their way into the air force by indirect routes. Eighteen-year-old Malcolm Beverly of Toronto had decided with a friend to join the navy, but on the appointed morning, the friend overslept. After waiting for him for about an hour, Beverly, "on the spur of the moment," decided to see about the air force, "and in particular, aircrew." He was accepted, and went on to pilot training, later flying Thunderbolt fighters. Frank Phripp of Toronto had also set out to join the navy with a friend, but no sooner had they entered the recruiting office than they both had second thoughts, turned around, and went straight to the air force recruiting office. Phripp completed a tour with Bomber Command, winning the DFC. He made the air force his career, retiring as a group captain in 1967. Tom Anderson of Ottawa was well connected with the army, having a number of uncles who were senior officers, including Major General T. V. Anderson, later chief of the defence staff. But shortly before the war, a family friend, Squadron Leader W. Kenny, took seventeen-year-old Tom for a ride in a Vickers Vedette flying boat. Thereafter, there was never the slightest doubt about which service Tom would join if war came.

Some recruits had to work hard to get into the air force. Wally Loucks grew up on an Indian reserve near Peterborough, Ontario. When he attempted to join the air force, he was told that he didn't have the necessary educational qualifications for aircrew. Undeterred, Loucks made his way to Queen's Park, home of the Ontario Legislature, and talked his way in to see his provincial representative. Impressed by the young man's enthusiasm, the MPP made some calls and paved the way for Loucks's entry. He became a wireless operator and flew a tour of operations with 419 Squadron. Jim Lovelace of Sydney, Nova Scotia, was equally determined to become an airman. He tried to join in January 1940 but the medical beat him. Lovelace was ordered to blow into a device called a respirometer, which measured lung capacity. He had to support a column of mercury with the power of his lungs. He couldn't. No matter how vigorously Lovelace puffed, the column of mercury refused absolutely to "stay put for the required number of seconds." He tried again in March 1940. The same thing happened. Three months later, Lovelace was back. This time, as the mercury column wobbled uncertainly, an airman burst in with the news of France's collapse. The MO said to Lovelace, "Well, I suppose they'll be needing you now!" and passed him as fit for aircrew. (Joffre Woolfenden also experienced difficulty with the respirometer when joining the air force in 1938; he made the mercury stay up by surreptitiously blocking the input tube with his tongue.) Ron Monkman of Toronto worked as a driver for a parcel delivery firm. One morning in June 1940, he and a fellow driver were informed that their loads wouldn't be ready for several hours – and, says Monkman, "if you didn't drive, you weren't paid." On an impulse, the two drivers decided to join up. Having no preference as to service, they agreed to go into the first recruiting office they came to on Bay Street. It was air force. Although his fellow driver was turned down, Monkman was accepted and became a photographic reconnaissance (PR) pilot.

Few of the BCATP's students had flown before enlisting. Mac Reilley of Montreal was one of the exceptions, although his air time was brief. One Sunday he and a friend "snuck out of Sunday school, took the streetcar out to Cartierville, and for five dollars went up for a short ride in an open-cockpit plane. We were thirteen years of age. As five dollars represented five weeks' work on the paper route, something about flying must have drawn me."

The politicians who had thrashed out the details of the BCATP felt, as politicians usually do, that they had all done a magnificent job. History would remember them. They posed for photographs in front of the Parliament Buildings, looking suitably statesmen-like (although Mackenzie King later claimed in his diary that such occasions were "the kind of thing that sicken me of public life"). [3] They had written their memoranda. They had drafted and corrected their press releases, polishing and blue-pencilling until they had achieved just the right tone. The reporters had conducted their interviews, jotted their notes, and filed their stories.

Now the time for rhetoric was over. Now the plan had to become a reality. And it was up to the RCAF to make it happen.

The diminutive force was ill-prepared for the task. Since its formation fifteen years earlier, the RCAF had been less an air force than a "national laboratory for almost every field of flight, in each of which it carried on until commercial operators entered the field, at which point the RCAF retired."[4] For years its small complement of non-threatening aircraft had been employed in such tasks as surveying, forest-fire spotting, exploring air-mail routes, photographing, and, from time to time, impressing the taxpayers by zooming around at air shows. Most RCAF officers were veterans of World War I, often described disparagingly as bush pilots in uniform. In the inter-war years, a grateful government had occasionally sent one of them to Britain to attend the RAF Staff College. On returning home, the lucky candidates would gamely try to implement the lessons they had learned in the defence of their homeland. It wasn't easy. Canadians refused to take the risk of air raids seriously. It was not as if the Americans would ever attack Canada. And if for some incredible reason they did, it would take more than a handful of low-powered, obsolete biplanes to stop them. Other potential enemies? They posed virtually no threat, everyone believed. They were all too far away. Flight Lieutenant (later Air Vice-Marshal) G. R. Howsam challenged this complacent attitude in *Canadian Defence Quarterly* early in 1931. He pointed out that aircraft carriers constituted a very real danger; they could sail to within a few miles of the Canadian coast, then send off attacking forces to bomb major cities. The only answer, he declared, was a Canadian bomber force that could bomb enemy carriers and destroy the enemy's bombers before they took to the air. At the time, Canada possessed no bombers worthy of the name.

As the *Official History of the RCAF* points out, the RCAF really didn't become a military force until the mid-1930s, and even then it was laughably small. In 1932, after a budget cut by the Conservative government of R. B. Bennett, the force could boast a total of

99 officers and 629 men. A few obsolete Siskins and Atlases (both products of the same British company, Armstrong Whitworth) constituted the offensive capability of the RCAF in those years. In 1936, the Wapiti appeared. Essentially a metal version of the venerable DH9A of Great War vintage, it incorporated all the shortcomings of the original and introduced a few new ones of its own. An open-cockpit biplane with a fixed gear, it came into service at a time when modern air forces were fielding monoplanes with enclosed cockpits and retractable undercarriages. Using the wings and tail unit of the ancient "Ninack," the Wapiti exemplified the make-do-and-mend approach to national defence of the period. RCAF airmen quickly dubbed it the "What a Pity." One of them recalled the Wapitis when they arrived: "The government had bought them for a dollar apiece from the British who had used them out in Afghanistan. They were armour-plated in some unique locations: there was a piece of boiler plate under the carburettor, and another piece strategically located under part of the pilot's seat. Those Afghans, by all accounts, were fantastic shots! The Wapitis, when they arrived, were full of sand, dirt and whatever. . . ."[5] Another, C. R. Dunlap, described the Wapiti as "a beast" that "glided like a brick."[6] The RAF's opinion was no more positive. Hamish Mahaddie, the notable bomber pilot and Path Finder of World War II, flew the Wapiti in Mesopotamia during the thirties as a sergeant pilot. "No one had a good word to say about it," he recalls.

Morale in the RCAF plummeted. Was there any future in the service? many wondered. It might have been a little easier to bear if the other services had taken cuts on the same scale as the air force. They hadn't. The air force seemed to be the target every time the government wanted to save a dollar or two. The government apparently regarded the air force as a kind of frivolity; nice to have to show off when times were good, but easily dispensed with when times became more difficult.

The rise of fascism changed all that. Uneasy about developments in Europe, the politicians came to the conclusion that perhaps the RCAF did matter after all. Slowly, the wheels of expansion began to grind. In 1935 thirty-eight candidates were enrolled in the air force, twenty-four of them winning their pilot's wings the following year. The RCAF ordered some two hundred modern aircraft and, reflecting the new awareness of the need for home defence units, planned to build airfields on the east and west coasts. As international tensions worsened, Canadians became increasingly worried by the possibility of air attacks.

Early in 1939, members of the House of Commons demanded to know what the government was doing about alleged moves by the Germans to establish seaplane bases in Iceland and, according to some rumours, in Hudson Bay. The government had to admit that at that time the maritime provinces' aerial protection was in the hands of two Stranraers and two Fairchilds; that of the West Coast depended upon five single-engined Sharks and eight other assorted types. New aircraft were on order but they hadn't arrived yet; these things took time, it was explained.

Six months before the outbreak of war in Europe, Mackenzie King budgeted no less than $60 million for national defence, about half of that amount being earmarked for the air force. At long last the permanent force began to assume slightly more impressive dimensions: 525 officers and 4,500 other ranks, with 220 officers and 2,172 other ranks in auxiliary units. During the Royal Tour in the summer of 1939, the air force was usually in evidence, providing escorts of the Siskins and Atlases that had been around for a generation. They made a lot of noise but impressed no one who had any knowledge of modern air power. Despite the gradual introduction of small numbers of Hurricanes and other new types, when war was declared most of the RCAF's units still flew ancient, unreliable equipment. During a trip to the East Coast to bolster defences, a gaggle of Wapitis of No. 3 Squadron landed in Maine after one of the flight commanders developed engine trouble. The members of his flight followed him in. It was a nail-biting moment in Ottawa. Would the United States impound the Wapitis? Cynics thought it unlikely, particularly if the Americans troubled to look the outmoded biplanes over. (The Wapitis were permitted to depart, and no doubt the Americans were glad to see them go.)

The rapidly expanding air force had a number of key responsibilities, including the manning and operation of the home defence squadrons and the various overseas squadrons as they came into being. But by far the biggest job it faced was the running of the BCATP, although so far the huge plan existed only on paper. The experts said that some forty thousand men and women, servicemen and civilians, would be required – instructors, maintenance staff, administrators, cooks, clerks, riggers, parachute packers, and heaven knows who else. *Forty thousand.* No matter how you calculated it, it was ten times the air force's current strength. The job would have been awesome enough if it had just been a matter of recruiting and outfitting forty thousand individuals. But this was a highly technical project, requiring experts to instruct, to service aircraft, and to administer. On top of that, the training plan involved the construction of about a hundred airfields and depots from one coast to the other. Only a handful of these fields already existed – Camp Borden, Trenton, Jericho Beach, Dartmouth, High River, Rockcliffe. Plus a few others and a score of flying-club fields and Trans-Canada Airlines (TCA) bases that could be put to use. The majority of the airfields for the nascent BCATP still had to be carved out of the earth, the buildings erected, the services installed.

Much credit is due the officials who made an early start on the whole complicated business. In October, before the negotiations that led to the creation of the BCATP had even begun, officials of the Department of Transport and the air force discussed the selection of airfield sites. They reasoned that, once the site selection was completed, engineering plans could be developed during the winter, with construction starting in the spring. An embryonic system of airports was already in place, the beginnings of the cross-Canada airline network that had been established before the war by the energetic minister of transport, Clarence Decatur Howe. In addition to the municipal airports, there were what was known as "100-mile" fields, located approximately every one hundred miles between the municipal fields in the thousand-mile stretch of bush and muskeg north of Lake Superior between North Bay and Winnipeg. With little additional work, these fields could, if needed, be quickly readied for training purposes.

The survey work began at once. Field work-parties, consisting of an inspector and an engineer from the Department of Transport and an RCAF officer, began the task of deciding where BCATP fields should be situated – and, equally important, where they shouldn't. In most cases, no detailed topographical maps existed and much of the basic survey work had to be done from the air. As a Department of Transport report stated:

> This procedure is not as simple as would at first appear, for without experience as a guide, any open space on the ground free of hills or gullies is apt to appear as a potential airport. Years of experience in this work had taught the investigators to beware of the lure of swamp and muskeg thereby avoiding much waste of time. Nevertheless, plotting these areas while in flight was a lengthy and arduous business; many of the areas were far removed from roads or even recognizable tracks and it was necessary not only to show the areas to be examined but also the best method of gaining access to them. One party on the prairies flew sixty hours in one month on the work alone. . . . Thereafter, the party moved by train or motor car into the locality where they could walk over and examine each site carefully in order to get full details regarding topography, the nature of the soil, drainage, availability of gravel, electricity, water, access to highways, railways, servicing facilities, etc. [7]

The survey parties looked for areas from three-quarters of a mile to one mile square, with a relatively uniform surface and a slope sufficient to ensure drainage but not exceeding .5 to 1 per cent. Good natural drainage was important because it obviated the need for deep ditches. Soil quality mattered too; it had to be sufficiently fertile to support a good sod, yet porous enough to provide adequate drainage. The absence of hills, tall buildings, trees, and telephone and power lines was clearly essential, as was the need for ample electrical power and water supplies. Additional requirements included the need for "the right quality of gravel adjacent to the site," reasonable proximity to the "social amenities given by towns and cities" [8] – and, ideally, the potential for civil use after the war. Unhappily, many of these requirements tended to be mutually exclusive. Soil producing good sod is seldom co-operative enough to provide adequate drainage, and a level or nearly level field, while it may be fine for flying, tends to retain water. The balancing of all these factors led to a "certain preference" [9] being given to southern Ontario and the southern portions of the prairie provinces. The survey parties made their reports to the Department of Transport. Any potential site that had the unanimous support of all members of a party earned a more detailed report, which went to the Aerodrome Development Committee for final approval.

Property owners pricked up their ears. Letters poured in to the Department of Transport offering land for airfields. Few were worth opening. "The writers of these letters rarely had the faintest idea of what airport requirements for the Air Training Plan actually were; and the net result was a considerable loss of time in looking at properties that in all but an infinitesimal percentage of cases had no potential value as airport sites." [10]

The fall survey – which probably got the BCATP started six months earlier than would otherwise have been possible – showed that twenty-four airfields were available immediately, needing only the addition of a few buildings to make them ready for training. Fifteen others needed more extensive work. About eighty would have to be built from scratch.

The Canadian National Railways real-estate people looked after the negotiations for the selected properties. They delighted few sellers; the government paid rock bottom prices. Traditionally, the Royal Canadian Engineers had taken care of construction work for the air force, but the scale of this project set it apart from anything that had been tackled in the past. Sensibly, the government's Air Council decided to form its own Directorate of Works and Buildings and selected an expert on major construction projects to head it. Dick Collard, the vice-president of a large Winnipeg engineering firm, Carter-Hall-Aldinger Construction, was the ideal man for the job, "one of the most capable and thoroughly experienced construction executives in this country."[11] In a matter of weeks, he had become a wing commander (later to be promoted to air vice-marshal), heading up the RCAF's own construction organization. He lost no time in hiring engineers and draughtsmen and plunging into the formidable job of designing the scores of buildings – the hangars, the barrack blocks, the offices, the messes and medical quarters – and arranging for their construction. To speed the whole process along, a high degree of standardization became a feature of the entire project, creating an efficient form of systems building in which standard components were used for buildings of different sizes and shapes. Similarly, the airfields and depots were built to standard patterns, depending on their end use; one layout for an EFTS, one for a SFTS, one for an AOS, and so on.

Dozens of important questions had to be answered before the work could begin on the sites, not the least of them being: Should the hangars for the BCATP's airfields be built of steel or wood? In peacetime, steel would have been the choice, but since the outbreak of war, all metals had become scarce. Use wood? Some experts doubted that wood trusses spanning a hundred feet or more were feasible. Nevertheless, the engineers developed a design using timber with a fire-resistant impregnation. The wartime Timber Controller ensured that the British Columbia fir industry produced the necessary timber. Tests soon proved the practicality of the design, and most of the BCATP's hangars were built using "heavy wooden trusses with bolted joints and supported by rigidly braced timber columns." [12]

The winter of 1939-40 was a frantic one for the air force staff in Ottawa. Group Captain E. C. Luke, who had been appointed to the post of deputy director of works and buildings, remembered it as "an everlasting pursuit of the impossible. . . . Time and date were ignored. Headquarters personnel toiled from nine in the morning until twelve at night, seven days a week." [13] Luke admits that the "heterogeneous organization" had its setbacks in the early days, but there gradually emerged an engineering and construction organization "which tackled, successfully, the largest undertaking of its kind in history." [14]

Throughout those hectic winter months, contractors' representatives and technical staff huddled in cramped meeting rooms and corridors of inadequate Ottawa offices in a fog of cigarette smoke, waiting with varying degrees of patience to see this or that official. There was seldom enough room to spread the big blueprints; they had to be examined, amended, approved, on knees or chairs. Mountains of correspondence wobbled on top of filing cabinets. Salesmen perched on their display cases; hats and coats hung from doorknobs. The constant clatter of comings and goings, the strained buzz of conversation, the jangle of telephones, each call more urgent than the last, all added to the sense of confusion and pressure. Time was short. Too much had to be done in too little time. And it had to be done right; if it wasn't, C. D. Howe, now the minister of supply (often known as the "minister of everything"), would demand to know why. The tireless Howe was one of the men who made the BCATP happen in record time. He never let red tape get in the way of getting the job done.

The new directorate had to struggle along in sadly inadequate quarters in the "Temporary Buildings" on Ottawa's Wellington Street. In spite of their difficulties in these early months, the staff produced more than 750,000 blueprints and 33,000 drawings. By the end of 1939, contracts began to flow. For No. 2 SFTS, Ottawa, a $267,869 contract went to Garvock Construction to construct accommodations buildings; $307,750 for 112-by-125-foot hangars was awarded to Brennan Construction; and $119,568 for the drill hall, hospital, and other buildings went to Dagenair Construction.[15]

On the sites, while winter did its best to discourage them, work gangs began the preparatory work, clearing the fields, levelling them, and pouring the foundations. Airfields had to be big enough to accommodate three landing strips between 2,500 and 3,000 feet long. Trucks streamed onto the airfield sites laden with wood, principally fir, pre-cut and ready for the construction of hangars, offices, storerooms, and all the other structures needed in a BCATP base. Local workers assembled the buildings, carpenters earning about seventy-five cents per hour, labourers forty-five cents. Many of the workers were inexperienced because they had not worked for years. They learned quickly. Soon they could assemble a hangar in a matter of hours, bolting the pre-cut components together, nailing the cladding to the walls, the shingles on the roofs, and hanging the ready-made doors. Much of the lumber had not fully dried out, and BCATP ground crews were destined to spend countless hours tightening and easing turnbuckles and nuts and bolts as the drying wood made the massive buildings stir like living things.

By early spring 1940, the first fields began to take shape, although most were short of such niceties as plumbing and heating, and roadways were strips of gravel perched uneasily on oozing, squelching mud the consistency of marshmallow. At that period, the BCATP fields resembled nothing more than a string of no-man's-land sets for World War I movies. Yet the work had to go on, and astonishing achievements were recorded. "In one instance," Luke recalled, "the keys to a finished school were handed to the RCAF opening party six weeks after the first truck had driven onto the farmer's field." But occasionally too much was attempted too quickly. "A party of VIPs made a tactical error in visiting a prairie school that had recently been opened in gumbo season. The inevitable happened. Their staff cars sank to the axles, and they were forced to carry on in high rubber boots, up to their knees in good old Manitoba mud. The next morning at 0901 hours an order emanated from a very high level to the effect that, at once, the very first thing to be done at every site was to pave the roads. It was," Luke commented, "a splendid but very impractical idea." [16] Later, some indefatigable statistician calculated that the runways and other paved surfaces on BCATP fields were the equivalent of a twenty-foot-wide highway running from Halifax to Vancouver. [17]

Water became a major problem on a number of sites, particularly in the prairie provinces. Notable examples were Moose Jaw and Mossbank, Saskatchewan, where the water was found to be highly mineralized. "The consequences of normal consumption are most distressing," [18] the students were warned with unwonted delicacy. Better water was found some seven miles away from Mossbank, but the distance meant that a pipeline and pumping system had to be built to get it to the school. Even then the problems of water supply were not over: the pipeline kept leaking because the wartime shortage of metal pipe necessitated the use of wooden staves bound together with hoops. Frequent drops in pressure resulted, a matter of more than a little concern to the unit's firefighters. (The story has a happy ending, for the problem eventually led to the installation of an indoor swimming pool at Mossbank as a means of storing water for fire-fighting purposes.) Water consumption at BCATP bases varied from twenty thousand to fifty thousand gallons per day, depending on the size of the school and its use. Bob McBey, an RAF trainee who arrived at No.

32 EFTS, Bowden, Alberta, early in 1941, was amazed to find that "there was no water on the station except what came in a cylindrical drum mounted on a 4-wheeled horse-drawn buggy converted to be pulled by a truck." This service version of the Depression-era "Bennett Buggy" continued to provide the base with water for several weeks.

Late in April 1940, the first BCATP students arrived at No. 1 Initial Training School (ITS), Toronto. To accommodate the school, the government had taken over the Eglinton Hunt Club on Avenue Road – in happier days a popular haunt of the country-club set. When the recruits arrived, the premises still boasted bowling alleys and a swimming pool, but sleeping quarters had to be set up in the stables – after a good deal of shovelling by the young airmen.

The first ITS course was unique, since 164 BCATP recruits joined fifty-seven pilot officers, the last of the prewar entrants. Seventeen of the intake washed out during training. Of those who graduated, ninety-three, some 42 per cent, lost their lives in action. They won three DSOs, and one Bar, thirty-four DFCs, and four Bars, one OBE, one MBE, and one AFM. Three of the students were Americans, the first of about 8,800 U.S. citizens who served in the RCAF during the war. [19]

While these students were at ITS, the RAF was suffering catastrophic losses in France. Outnumbered and often outmatched, British airmen were shot down in appalling numbers. Some bomber and army co-operation squadrons lost their entire strength in ill-conceived sorties against the supremely confident Germans. Would the unthinkable happen? Would France *fall*? If it did, how long before the Germans would invade England? And how would they be stopped? To many it seemed that this was the time to shore up Britain's defences with every available aircraft and pilot – including instructors and students in the final stages of their training. Perhaps the BCATP should be put on hold until the situation improved.

The British refused to panic, declaring that the "efficient prosecution of the war can best be achieved by adhering to the plans laid down for the Air Training Scheme and by accelerating them to the utmost." [20] It was a heartening response, but the notion of actually *speeding up* the plan shook the BCATP to its insubstantial foundations. "Who can forget the staggering announcement from the CAS [chief of the air staff] in May 1940," wrote Luke. The air force demanded that construction originally scheduled to be completed in two years should now be done in only one. Luke described it as "sheer madness." [21] The powers that be must have taken leave of their senses. After a particularly harrowing day, one man at air force HQ in Ottawa declared, "If this damned thing works out it will be one of the greatest coincidences of all time." [22] He had a point, for the whole vast project consisted of countless other projects all coming together at the right time at the right place: airfields, hangars, aircraft, engines, tens of thousands of spare parts, personnel, vehicles, and all the multitudinous bits and pieces, specialist skills and services needed in a training establishment.

The lack of any one element could throw a monstrous wrench into the works. And did. When No. 1 EFTS at Malton (now Toronto International Airport) opened for business in June 1940, it was discovered that the air force had sent tools and spare parts for Fleet Finch trainers, despite the conspicuous presence of rows of Tiger Moths out on the tarmac. When crated Ansons and Battles arrived from Britain, they couldn't be assembled, because no one had thought to provide the right nuts and bolts (the threads of British and North American fasteners were not compatible). BCATP ground staff had to make their own wrenches and other tools, and, in some cases, parts for the aircraft.

They discovered depths of innovative talent they hadn't realized they possessed.

In the early summer of 1940, anticipating an invasion at any moment and desperately marshalling every weapon at their disposal, the British put a stop to the export of Ansons and Battles. One ship, its hold full of crated Ansons, was already on the high seas at the time, headed for Canada. It had to turn about and return to port. Observers shook their heads in dismay. Britain's plight must be desperate indeed if the defence of the Mother Country depended upon such outdated and inefficient aircraft. The Anson was a twin-engined reconnaissance bomber, well past its prime but admirably suited to the task of training pilots, observers, and wireless operators. The Battle, a single-engined light bomber, had suffered ghastly casualties during the German advance through France and had been withdrawn from front-line service. It, too, was destined for training duties in the skies of Canada.

Only relatively few of these types had so far been delivered to Canada. And now, with the situation in Europe, there was no telling when – or even whether – any more would be available. Could the Anson be manufactured in Canada? Everyone nodded. Canadian firms were already fabricating parts for the "Annie," so there seemed no reason why major components shouldn't be tackled. And while they were at it, the factory could incorporate a few modifications to make the airplane even better suited to its work as a trainer operating in Canada. Improved heating and insulation and more cabin space ranked high among the projected upgrades. Engines posed a problem, however. Canada had no aero-engine industry, and the British wouldn't release any more of the Cheetah engines that powered the Anson. In his damn-the-torpedoes manner, Howe plunged into the task of replacing the Cheetah. He found a likely prospect in Pottstown, Pennsylvania, where the Jacobs company produced an engine known as the L6MB. Although not as powerful as the Cheetah, the Jacobs was

an acceptable substitute. Within hours, Howe's staff had negotiated an order for more than two thousand L6MBs (the government's New York bankers signing the cheque, since Jacobs was
dubious about Canada's credit in that tumultuous year). At the
same time, the lawyers busied themselves with the creation of a
new Crown corporation, Federal Aircraft of Montreal, to
assemble Canadian Ansons. The company's initial order was for
one thousand Ansons, the first to be delivered in February 1941.
Unfortunately, the program ran into endless delays, with delivery
of the first aircraft nearly a year late. Although the Anson II incorporated several improvements over the original British version,
its Jacobs engines made it underpowered, and it was used only for
pilot training, since the aircraft employed on navigation exercises
usually carried three or four airmen. A fully loaded Anson II with
wooden propellers was incapable of staying aloft on one engine.
In all, 1,832 Anson IIs were built. Several improved versions came
along to replace the early Ansons, but many were still in service
when the BCATP came to an end in 1945. [23]

The historian Leslie Roberts wrote that "what had been Britain's Plan, carried out on Canadian soil, now became Canada's
Plan, pure and simple. Originally the Plan had been geared to
Britain's programme of aircraft production and would turn out
graduates according to the number of aircraft Britain would be
able to supply. . . . That the Plan survived at all, that it came into
flower months ahead of original schedules, is one of the miracles
of the war." [24]

Although Canada was woefully unprepared to take on the responsibility of running the BCATP, in one respect the nation did have
an aerial ace up its sleeve: the Canadian Flying Clubs Association.
Formed in the twenties by the civil aviation branch of the Department of National Defence to encourage the development of civil
aviation, the association helped to establish twenty-six flying

clubs in various parts of the country. Four failed, but the remaining twenty-two survived and were still active when war broke out. President of the association was Murton Adams Seymour, a lawyer and former RFC pilot. Seymour had learned to fly in 1915 and had founded the nation's first flying club in Vancouver. Early in 1939, he approached Ian Mackenzie, the minister of national defence, suggesting that in the event of war the personnel of the flying clubs could train air force pilots up to the elementary level. Mackenzie liked the idea. In June, eight clubs began training. Although the air force chiefs were far from happy about civilians instructing their pilots, they had no better suggestions. Now known as Flying Training Companies, the clubs became limited liability companies. To demonstrate "financial stability, community support and good faith," each had to raise $35,000 working capital locally and satisfy the RCAF as to the competence of its instructors, administrators, and maintenance staff.

The hierarchy of the typical civilian BCATP school was complicated. A civilian manager, with the status of a group captain, was in overall charge of the school. At the same time, a supervisory RCAF officer was responsible for the quantity and quality of the graduates. The aircraft belonged to the air force, but in all other respects the schools were civilian operations. Elementary Flying Training Schools were numbered: No. 1 at Malton, near Toronto; No. 2 at Fort William, Ontario; No. 3 at London, Ontario; and so on, a string of schools extending from Stanley, Nova Scotia, (No. 17 EFTS) to Boundary Bay, British Columbia (No. 18 EFTS). The only exceptions to the club rule were the schools at Cap de la Madeleine, Quebec, and Davidson, Saskatchewan. The former was run by Quebec Airways, the latter by the RCAF.

The vast influx of students created an instant shortage of flying instructors. An endangered species during the Depression years, instructors suddenly found themselves in demand. To meet the need, hundreds of experienced civilian pilots, many of them former bush pilots, and a lot of them Americans, were recruited. Given brief tests – which quickly weeded out the inevitable hopefuls who had claimed far more flight time than they actually possessed – they went through training sessions at a flying instructors' school (FIS) before going to an elementary training school and acquiring their first students. Soon these instructors were joined, albeit reluctantly, by the first pilots to graduate from the BCATP. Almost all of these graduates became instructors, much to their dismay. Civilian pilots also played a major role in the training of observers and wireless operator/air gunners.

Canadian Pacific and a number of smaller firms agreed to take over control of ten air observer schools, with much the same sort of financial arrangements that had been worked out with the flying clubs. In this case, however, instruction was provided by RCAF personnel. The companies' civilian staff pilots flew the aircraft, principally the Anson and the Norseman, on navigation and radio instruction exercises respectively; many civilian pilots also flew Battles and Lysanders used in gunnery and bombing training.

Had they been offered a choice, most of the American pilots working for the BCATP would undoubtedly have preferred to be working for a U.S. airline or in the U.S. Army Air Corps. But until the Japanese attacked Pearl Harbor in December 1941, the United States was still at peace, and flying jobs were far from plentiful. If you were married, under twenty or over twenty-six, or lacked at least two years of college, the doors to the Army Air Corps remained locked and bolted. While some Americans stayed for only a short time in the BCATP, some seven hundred worked for years beside their Canadian colleagues, instructing or performing the duties of staff pilots. It's hard to imagine the BCATP being able to expand as rapidly as it did without the contribution of the civilian flyers. They were indispensable in the early days.

Although the thought of flying training invariably conjures up images of fresh-faced young *pilot* trainees at work, the BCATP had been organized to train observer/navigators and wireless operator/air gunners too (although neither aircrew trade existed in the RCAF before the war). Later, bomb aimers, flight engineers, and other specialist aircrew trades would be added to the list. In fact, more observers, air gunners, *et al.,* went through the system than pilots: some 81,000 versus slightly more than 49,000.

From the start, the needs of the RAF shaped the plan, just as they shaped the training plans in Rhodesia, South Africa, and other parts of the Commonwealth. The idea was, no matter where an airman had been trained, he would eventually bring a standard level of skill and knowledge to his squadron and would be inter-changeable with any other product of the scheme.

The administration of the BCATP was becoming an increasingly onerous and complex task. Norman Rogers, the minister of national defence, decided that it should become the responsibility of a cabinet minister. He chose well. Charles Gavan Power – known as "Chubby," although he was not particularly rotund – was an affable politician who would earn the respect and affection of his air force colleagues and contribute much to the success of the BCATP.

Another appointment of great significance took place early in 1940. RCAF HQ asked London to nominate a suitable officer as director of training in Ottawa. London's choice was Robert Leckie, the man who had first broached the concept that became the BCATP. At the time, Leckie was in charge of the RAF in Malta. Although prepared to leave for his new duties immediately, Leckie discovered that his appointment had ruffled a number of important feathers in Ottawa. The reason: he was senior to the officers to whom he was supposed to report, notably Air Commodore G. O. Johnson, then in charge of organization and training at RCAF

HQ. In fact, Leckie was senior to every officer in the RCAF with the exception of the chief of the air staff, George Croil. One might think that the air force had more important things to worry about at that particular juncture in history, but Croil was so concerned that he cabled London, requesting someone other than Leckie. London said no. Characteristically, Leckie ignored the commotion. When he arrived in Ottawa early in February 1940, he set to work quite unconcerned about reporting to Johnson, a junior in rank, years of service, and age.

Leckie's position was far from easy, for he had to balance the demands of both the RAF and RCAF. The short, wiry – and sometimes devastatingly direct – Leckie demanded much from his staff and had no patience with incompetence at any level. His colleagues liked him, though they were wary of his bristly temper. Leckie would become a major influence on the BCATP as it developed into one of the largest air training organizations in the world.

Three months after Leckie took up his new post in Ottawa, the first BCATP students were about to embark on the lengthy process of learning to fly. One of them was John Simpson of Kingston, Ontario. Employed as an apprentice air engineer at the Kingston Flying Club, he possessed a private pilot's licence and had already been accepted by the RAF a few months before the outbreak of war. In July 1939 he had been instructed to travel to Britain at his own expense. "While I was trying to find a way over," he says, "I got a cable in early August, telling me not to come, because of the international situation, but to apply to the RCAF instead." Two days after Canada's declaration of war on Germany, Simpson hitchhiked to Trenton and volunteered to serve as a pilot. To his dismay he found that peacetime restrictions still applied. He didn't qualify for aircrew in the RCAF, even though he had a pilot's licence and the RAF had already accepted him. "I was told that I must have an engineering degree to enlist as a pilot, but I would be accepted as an aero-engine or airframe mechanic." Simpson told the recruiting officer that it had to be pilot or nothing, and stalked out. Another six months elapsed before the air force decided that perhaps he was worth considering after all.

All air force careers began in cheerless edifices called manning depots, vast and unlovely places in which thousands upon thousands of bunks had been crammed into an area that looked large enough to house the *Graf Zeppelin*. So numerous were the bunks that a visit to the bathroom during the night was likely to lead to an embarrassing and sometimes incredibly lengthy ordeal of trying to find the right bed on one's return. It was a world of lines – for food, for clothing, for bedding, for shots, for pay, for a haircut, for a basin to wash and shave in. Young men from every province, every state, from every walk of life, coped as best they could with the abrupt transition to service life. Quite apart from the so-so food, the bleak surroundings, the endless waiting, there was the overwhelming presence of others – strangers, individuals who suddenly became intimates, sharing every aspect of life, and whose personal habits could be unappealing. Ken Fulton, who became a navigator with 426 Squadron, encountered a recruit at manning depot who never showered and whose body odour "became unbearable." His comrades unceremoniously tossed the malodorous one into the shower. For many recruits, lack of privacy was the worst aspect of service. For others it was the merciless uniformity. Individual preferences counted for nothing. Bodies were there to be trained and worked until they resembled airmen.

The air force provided all the essentials of life: coarse air-force-blue uniforms and hefty boots; four shirts, two pairs of underwear, socks, the indispensable "housewife" (a small kit containing needle and thread and spare buttons); and a regular-as-clockwork $1.30 a day into the bargain. To some it was roughing it, to others it was comparative luxury; it all depended on how civilian life had been treating you. Frank Covert, later a navigator in Bomber Command, had been a lawyer in civilian life. When he lined up for his first air force meal, he told the server handling the gravy ladle that he didn't care for gravy. This seemed to amuse the man, for he made a scornful remark about "à la carte" orders and slapped two brimming helpings on Covert's plate. Ron Cassels recalls how upset one recruit was when he found he had to make do with only one towel for a week: "He did not think he should have to dry his privates and his face with the same towel." John Simpson remembers being "overwhelmed by the vast numbers of men from all over Canada, about 1,500 to 2,000, housed in this huge barn of a room [the Coliseum at the CNE grounds, Toronto] in double-decker bunks." The building had previously held horses and cattle, and the floor of the "bull pen," where the recruits drilled, consisted of several strata of manure. "It was watered every morning," Simpson recalls, "but it dried out as the day progressed and a layer of dust arose from the hundreds of marching feet and was inhaled. After ten days or so we all developed a sort of hack which was known as Manning Pool Cough."

The month or so at manning depot was nothing if not eventful. The air force packed every day with activity: an hour or two getting bodies in shape with knee bends, toe-touching, and running on the spot; endless thudding around the parade ground, learning to march with the puffed-up, shoulder-high swing the air force demanded; plus interminable lectures, shots, cleaning and scrubbing, painting and polishing. It was all part of the process of acquiring the never-quite-defined quality of "airmanship." The recruits had other names for it. The regimen affected men differently. Arthur Angus, an optometrist in civilian life, lost thirty pounds at manning depot. "It was good for me," he admits. On the other hand, Frank Murphy of Toronto *gained* thirty pounds. Most recruits found their days at manning depot more than a little disillusioning; the air force seemed to go out of its way to dampen any enthusiasm that might be possessed by its newest members. When Jim Emmerson of Georgetown, Ontario, was shipped to a depot at Lachine, Quebec, he was aghast:

> I had been innocent and naïve enough to believe the recruiting posters, which showed a smartly dressed young airman smiling his way into the hearts of all females. . . . As we marched from the Lachine railway station to the manning pool back gate, I saw a motley crowd of head-sheared people in dirty, slovenly khaki dress engaged in some dirty clean-up chores on the grounds. They all shouted, "You'll be sorry," the standard greeting to newcomers, I was to learn. I couldn't bring myself to believe that these drudges, who looked liked convicts, were actually airmen bent on the same plan as myself.

John Turnbull came from the tiny town of Govan, Saskatchewan, and was one of three brothers who served as RCAF aircrew during the war. He remembers manning depot as the place "where we learned not to volunteer information – such as in response to the Sgt. Major's query, 'Who're here as pilots?' His follow-up [with a nod to a nearby trash heap] was, 'OK, you pile it here and you pile it there!'"

Although every waking moment was shared with countless others, life at manning depot could be extraordinarily lonely, particularly for those who had never been away from home before. No one gave a damn about you except to make sure you obeyed orders. You didn't make any decisions. Life narrowed to its essentials: sleeping, eating, and surviving – getting over the next unpleasantness so that you would be ready to face the one after that. As far as the air force was concerned, you were a cipher, something to be processed. Air force recruit Alex McAlister remembered being herded from room to room where "uniforms" looked down throats, took pictures and fingerprints, and asked endless questions. He soon received his uniform, which didn't fit.

His boots were worse; too big for his feet, they flopped and slapped the floor when he walked. When McAlister visited the barber, he was pleased to be asked if he wanted his sideburns. "This, I felt, was very thoughtful of him, considering the number of jobs he did per day, and naturally I answered 'Yes' emphatically.

"'Well, catch 'em,' he sneered as he sheared."[25]

Shots affected some men violently. Glenn Bassett had five inoculations at No. 2 Manning Depot, Brandon, Manitoba, in the oppressive heat of July 1941, after which he had to march to the accommodations, an old warehouse half a dozen blocks away. When Bassett arrived at his bunk, he collapsed face down, unconscious. He was still there the following morning – and was charged with being late on parade.

Eventually uniformed, inoculated, and vaccinated, checked and double checked as all assembly-line products are supposed to be, the recruits were ready for the next stage of the process. But in nine cases out of ten, the next stage was not ready for them. They had to wait their turn until those further down the line had completed their courses and moved on. Since the assembly line could accept only so many at a time, something had to be done with those awaiting placement. The air force had the answer, or rather two answers: tarmac duty or guard duty. The former was a typical service euphemism for airfield labourer. You were available for any duty that the NCOs in charge elected to toss your way. You might be washing Tiger Moths in the morning, sweeping the hangars in the afternoon. Whatever needed doing and involved the minimum of intelligence and ability, you did, for you were an untrained aircraftman second class, an AC2, the lowest form of life in the air force. Jim Emmerson recalls being sent to an equipment depot in Montreal. "The station commander welcomed us with the heartwarming news that here at his equipment depot we were about to begin our great adventure with aircraft. 'Why,' he announced glowingly, 'you will actually get to handle the various parts of the aircraft.' That we did," Emmerson notes, "counting thousands of nuts, bolts, and unrecognizable things as inventory was taken on the station."

Work at airfields introduced many recruits to the dark side of air force life. On his second day of guard duty at No. 5 SFTS, Brantford, Ontario, Steve Puskas of nearby Hamilton was assigned to guard a crashed Anson. He arrived at the crash site as the bodies were being removed, and he recalls that "the seriousness of this business started to sink in."

Dave McIntosh of Montreal went to Toronto and became involved in "months of guard duty at Mimico racetrack. There was nothing to guard, of course. It was a place to put people for a time – part of the human resources pipeline as one would say today."[26] Allen Caine of Toronto was sent to Camp Borden for guard duty, expecting to be there six or seven weeks, but the system developed a glitch, and Caine found himself there for eighteen tedious weeks. Harry Holland experienced a lengthy spell – or rather several spells – at assorted time-occupying duties: two months at Paulson, Manitoba, a month at Moose Jaw, Saskatchewan, followed by a month at Saskatoon. "As we later learned," he says, "there was an almost complete breakdown in the 'pipeline' and some of us were just unfortunate to be caught up in it." Tom Anderson found himself on guard duty at Prince Rupert, patrolling the west coast "on the lookout for boarding parties from Japanese or German submarines." The airmen carried elderly Lee-Enfield rifles, which would have been of little use against enemy invaders, since none had breech blocks.

A few lucky souls sidestepped the tiresome guard-or-tarmac stage by being chosen for one of the air force's precision squads – "a select, finely honed flight, resplendent in white belts, gloves, and rifles . . . sent to wave the RCAF flag across the country,"[27] in the words of Norman Shrive. Shrive, who came from Hamilton, Ontario, was selected for the squad on a Friday morning in February 1942 – and was out of it the same day, after suffering minor injuries in a road accident. The RCAF's precision squads were

favourites at bond rallies, major sports events, and other functions, executing complicated sequences of parade-ground drill
without a single word of command.

When the human production line had room for you, you progressed to initial training school. You had already been selected as
likely material for aircrew, but at ITS the air force satisfied itself
that the earlier decision had been correct. Of even more importance, at ITS it was decided what *form* of aircrew you would
become. In the early days it was a question of pilot or observer;
those selected as wireless operators and air gunners usually went
straight from the manning depot to their schools. But there were
the inevitable exceptions. Howard Hewer of Toronto was at ITS in
the late summer of 1940. He recalls, "We were all paraded in the
drill hall to hear an address by Group Captain Geoffrey O'Brien.
After some platitudes about how he would wish to be a wireless
operator/air gunner if he were just starting out, he announced
that a new wireless school was just opening in Calgary." A
moment later the blow fell: the entire course had been posted to
the new school, putting paid to many a fondly held dream of fame
as a fighter or bomber pilot or an ace observer.

At ITS the air force sought to eliminate as many of the inevitable "borderline cases" as possible before they became involved in
flying training. ITS was tough and concentrated, with lectures
and tests on such subjects as navigation, meteorology, air force
administration, airmanship, aircraft recognition, and theory of
flight. There was an overwhelming sense of being watched, studied like a specimen in a laboratory, as you went about your duties
– and from time to time a member of your class would disappear,
"washed out" because he didn't measure up in some way. On
average, about 12 per cent of students failed ITS. [28]

They had all sorts of ways of testing you. The decompression
chamber, for one. It looked like a factory boiler; you sat in it while

the technicians fiddled with their gauges to adjust the pressure inside. They wanted to find out how well you could tolerate a lack of oxygen. You soon found that it did peculiar things to you, including playing games with your sense of colour. Purples started to take over: first your fingernails, then your neighbour's lips; at the same time everyone around you looked increasingly stupid, while, to your dismay, you found the simplest problem in arithmetic had suddenly become as taxing as the theory of relativity. Then they measured your brain waves under the electroencephalograph, because there was supposed to be a correlation between high EEG readings and failures in flying training.

If there was anything wrong with you, the medical officers at ITS would find it. Murray Peden of Winnipeg attended ITS at the University of Alberta, Edmonton. He describes the medical there, known as M2:

> [It was] a four-hour ordeal replete with careful colour-vision and depth of vision checks, and odd items such as testing a man's ability to balance on one leg with his eyes shut. During the first two and a half weeks of the course, while the M2s were going on, people were frequently to be seen in their rooms teetering on one leg, eyes shut and arms outstretched, as they practiced for this balance test. Since we were subsequently taught to disregard our unreliable senses completely and place our trust in instruments when visibility was cut off, I often suspected the Air Force of an inconsistency here – but I suppose one would have doubts about a person whose balance was so bad that he pitched onto his head the moment he closed his eyes, or who regularly soaked the linoleum with a coffee cup carried at a 45-degree angle. [29]

Norman Shrive, who had lost a place in the precision squad, recalls that one part of the process at No. 1 ITS, Toronto, was a visit to the psychiatrist – "a word that in 1942 still had connotations of voodoo and witchcraft." Shrive was lucky; as he entered the psychiatrist's office, a friend was just leaving. Passing Shrive, he whispered, "Get on the bus, for God's sake." Puzzled, Shrive went on in to see the psychiatrist. Within a few minutes the meaning of his friend's cryptic hissings became clear. The "shrink" asked Shrive if he pressed his trousers before going on a date. Shrive said yes.

"After walking to the bus stop – the bus in sight – did you ever say to yourself, 'Oh, God, did I remember to unplug the iron?'"

"Yes, sir."

"What did you do?"

Shrive writes, "No matter what I *really* would have done, I gave the required answer; the smile on the guru's face confirmed that I had pressed the right button. Pilots *always* got on the bus."[30]

The air force had seven ITSs in various parts of the country: Toronto (Nos. 1 and 6), Regina (No. 2), Victoriaville (No. 3), Edmonton (No. 4), Belleville (No. 5), and Saskatoon (No. 7). All but the brightest of students found it necessary to spend much of their free time studying to keep up with the avalanche of information that the air force kept sending their way. Algebra and trigonometry soon became the most detested subjects, for they caused the largest number of failures among students. The ambition of most students was to do well enough in math to pass as a pilot trainee but not to do *too* well, because it was widely believed that those who excelled were automatically recommended for training as observers.

At ITS, budding pilots finally got to sit in a cockpit, although not in a real aircraft but in a claustrophobic little torture chamber known as a Link trainer. The forerunner of today's flight simulators, the Link became as feared and loathed among many of the trainees as algebra and trigonometry, and for the same reason. The merciless thing could end your career as a pilot before it had started. If you failed to co-ordinate its tetchy controls within a minute or two, you were out, with little hope of another chance.

Curiously, some trainees did well on the Link, only to fail when they took the controls of a real aircraft.

The idea behind the Link was sound enough: the authorities could obtain a reasonably accurate appreciation of a student's potential without having to go to the trouble and expense of taking him flying. What made countless students loathe the Link was its extreme sensitivity. Wilfully unstable, it seemed to be balanced on a pinpoint; its stick and rudder had to be worked ceaselessly to keep the machine on anything approaching an even keel. The Link room itself added to the student's discomposure. Alex McAlister was amazed when he found himself in a circular room, complete with a domed ceiling representing the sky. He clambered into the Link, finding it tricky to tuck his legs into the tiny cockpit.

> The next thing I knew I was diving madly earthward and hadn't the faintest notion what to do about it. All previous instructions did a hasty exit from my dizzy brain, and the one thing to do was to grab the only available object which proved to be the "stick." It worked! In fact it worked too well! It became apparent that you don't need football tactics in flying. A mere push of the stick to the left sent the left wing down and a tug toward you brought you climbing into the clouds. The instructor kept shouting into the intercom, and, although I heard every word he said, he must have thought I was as deaf as a stone. "Left rudder pedal causes left turn. Right rudder pedal causes right turn!" he bellowed. "So what?" I answered silently. After all, it's one thing to tell a person to juggle twelve golden balls, but another thing . . . to do it yourself. Finally, when the instructor appeared completely exhausted, he pronounced the test over, and I emerged from the "aircraft" feeling as if I were ten years old coming off the "crack the whip" at the midway.[31]

Frank Murphy of Toronto also found the Link a capricious contraption. During his test, he received the order, "Turn to port," through his earphones. Momentarily confused, Murphy turned to starboard, hit the instructor with the Link's stubby wing, and knocked him off his feet. As the instructor dusted himself off, he said, "You'll make a great WAG, Murphy." Murphy did.

They decided your fate at ITS. Were you pilot material? Or would you be better employed as an observer? Your entire air force future hung in the balance. Yet the authorities seemed to be in no hurry to make up their minds about you. The tests groaned on, checks galore of your knowledge, your skills – particularly math – your co-ordination, your senses, your reactions, your very being. The authorities seemed convinced that they'd find something wrong if only they looked hard enough and often enough. The whole nerve-clenching process culminated in the aircrew selection board, the few minutes that usually resulted in the final decision about your air force career. The two or three officers comprising the board did their gallant best, but they never succeeded in putting anyone at ease. It was a sink-or-swim moment, and everyone involved knew it. In the mass-production climate of the BCATP, no time could be wasted on those who failed to make the right impression the instant they walked in. Did the flight lieutenant on the right have a hint of a frown on his forehead? Wasn't that a yawn the squadron leader with the DFC just stifled? Did they want to hear that you hoped to become a pilot because you wanted to fly, or because you felt it was the way you could best serve your country? Or was that overdoing the patriotism angle a bit? Countless students rehearsed their speeches while waiting to go before the board, quietly going over and over the words, correcting and smoothing until the last instant – only to forget every meticulously crafted phrase in the tension of the moment. To this day there are men of advanced years who still regret what they said or how they acted before their aircrew selection boards half a century ago. If they had mentioned this or avoided that, things might have been so different . . .

Nine out of ten students prayed to be selected for pilot training. It was the ultimate, the apogee of their ambitions, proof that they had been rated as potentially good enough to wear the coveted double wing on their tunics. Murray Peden recalled the stress he and his friend Gort Strecker felt as the moment of truth approached:

> I lined up, hardly able to breath. Flying Officer Milson began to shout out names, in alphabetical order. The first man's name was called: "LAC Baker . . . Observer . . . Edmonton AOS." I almost choked. On he went, the next two picked as Pilots and sent to No. 5 Elementary Flying Training School (EFTS) at High River. Then more Observers; back and forth. Suddenly he reached the P's and I stopped breathing: "LAC Peden . . . Pilot . . . High River." I was elated. I broke into a grin, then remembered guiltily that he hadn't reached the S's. In a moment it came: "LAC Strecker . . . Observer . . . Edmonton." I walked over to him, ashamed at my own joy, and tried to think of something to say. His eyes were swimming. "Ah, hell, I didn't want to be a Pilot anyway," he said brokenly. Neither of us could say anything more for a while; the blow was too heavy. [32]

At Alex McAlister's ITS they announced the students' postings on the notice board. With "pounding heart and knotty nerves," McAlister searched the "Pilot" list for his name.

> The terrible physical and mental state was intensified by watching those ahead of you as they received the tidings of their fate. . . . The "Pilot" list became clearer, clearer, clearer. Suddenly, without warning, it was there – right

before my eyes! Tears of joy blurred the seething mass of blue uniforms about me. The powers that be had chosen *me* to be a pilot! [33]

It is not without significance that both these accounts mention tears, an indication of how intensely trainees felt about becoming pilots. Those selected knew they were in for a long and demanding series of courses – from elementary flying school to service flying school, thence to operational training or instructor school. Many months – perhaps even years – would elapse before they were fully trained and ready for fighters or bombers or whatever the air force had in store for them. It would be a time of pressure and challenge. Many would fail. Washouts regularly eliminated one in three students, sometimes polishing off half a course, according to the know-it-alls who always seemed to be in evidence at such times.

When they finished ITS, the students had overcome the first hurdle in their training. They had been tested and graded; the experts had made their decisions. Proudly, the students sewed the propeller insignia of the leading aircraftman (LAC) on their sleeves, and they contemplated an increase in their earnings to $1.50 per day plus 75 cents flying bonus as they slipped the white flash of the aircrew trainee into the fold of their caps. (These flashes caused some confusion initially. Howard Hewer of Toronto recalls arriving at Calgary as one of the first students at No. 2 Wireless School: "Because of our white aircrew cap flashes, army privates thought for a while that we were all officers, and were saluting us downtown!") Wearers of the flashes chose to ignore stories that the army and the navy – and even RCAF ground crews – were telling local girls that the flashes identified VD cases. At last the serious business of learning to fly was about to begin!

CHAPTER THREE

Pilot Training: The Beginning

Fleet Finch

"Pilots will not wear spurs while flying"
– official instructions to airmen, 1918

They promised to keep in touch, but few did. ITS had seemed a momentous event in their lives, but now they were on their way to flying training school, and suddenly what had gone before dimmed into insignificance. They couldn't wait for the flying to begin. Unfortunately, in countless cases, they had to. The frequent blockages of the BCATP pipeline made it inevitable. At No. 2 ITS, Regina, fifty pilot trainees drew straws for the five vacancies available at No. 8 EFTS, Vancouver, also known as Sea Island. An amiable young Albertan named Rayne Schultz was one of the lucky ones. A fellow student offered him a hundred dollars to take his place. Schultz turned the offer down.

The EFTSs were real airfields, most of them stuck out in the wilds, umpteen windswept miles from any form of fun. Not that there'd be time for any fun, the students were told; they'd be far too busy. There were aircraft to be seen, little trainers, yellow jobs, buzzing about, taking-off, landing, taking-off, landing. And good food was to be found at most civilian-run elementary flying training schools. After the institutional fare of the manning depot and ITS, it was like taking up residence in the dining room of the Chateau Laurier.

Most former students remember their EFTSs with affection. Norman Shrive – who had assured the psychiatrist at ITS that he could be relied upon always to get on the bus – trained at No. 12 EFTS, Goderich, Ontario. He says without hesitation that "the eight weeks that began on an August evening when Milt Sills and I arrived at the little flying field perched on the edge of Lake Huron were, I am positive, the happiest of our lives." [1] Bill Martin of Bristol, England, a Fleet Air Arm trainee who, like Shrive, learned to fly at Goderich, concurs. It was, he says,

> Nirvana for eighteen- and nineteen-year-olds. . . . It was the age of bobby soxers and sloppy joes. It was soda fountains and malted milks and Big Band time on the juke box. Glenn Miller vied with Artie Shaw, Bing Crosby with Frank Sinatra. . . . It was Andy Hardy time. And we were teenagers and servicemen let loose on a continent worshipping both. [2]

In contrast, Al Forbes of Drumheller, Alberta, had less positive recollections of No. 5 EFTS, High River. A corporal met the new draft of students and marched them around, helpfully pointing out the location of the school's facilities. At the conclusion of the tour, he led his charges into a hall where the station commander and other senior staff members delivered suitable words of welcome with the usual urgings to study hard and play the game. "The others had left when the station padre showed up to speak to us about our spiritual life and to assure us that his office was there to help us. As the padre finished," says Forbes, "the corporal jumped up on the stage and asked for three cheers. We responded enthusiastically in honour of a man we had known for no more than ten minutes." When the padre had thanked the assembled students and departed, the corporal took over the stage again. The padre, he declared, was "a prince of a guy" – but, he added, his features becoming a study in lugubriousness and his voice dropping to sombre tones as if announcing the man's imminent demise, the padre was about to be posted away from High River. Wouldn't it be great to give him a going-away gift from all the students? By good fortune, the corporal said, he had a forty-eight-hour pass which he intended to spend in Calgary. He could pick up a suitable gift for the padre. "That said," recalls Forbes, "he asked two of our mates in the front row to pass their hats to raise money for a little gift. He ostentatiously dropped a few bills in the first hat to get things going." Forbes adds, "Many of us were aghast at this performance, particularly when we saw the generosity of our fellow LACs. We never saw the corporal again. Someone heard that he was later caught and charged. We thought of our vanished funds as the price of our first 'lesson' at EFTS."

Most student pilots lost little time in struggling into their newly issued flying gear: leather helmet and goggles straight out of *Hell's Angels,* a two-piece flying suit complete with fur collar, a bevy of pockets secured by a mile or two of zip fasteners, fleece-lined flying boots, hefty gauntlet gloves. It took a while to sort out the outfit's complexities, but at last you had it all on. It felt good and businesslike; you hardly noticed the rivers of sweat running down your legs as you studied the reflection in the mirror. Was that really you? That intrepid aviator? Or was he a polar explorer?

Before you had been there more than an hour or two, you had heard the *gen* from more experienced students who had done it all and seen it all and couldn't wait to tell you all about it. About LAC Plonk, who turned too tightly and spun into a farm half a mile from the main gate. How they were still finding bits of him in the back yard. About the *vicious* winds that could – and invariably did – spring up *at a moment's notice* in this particular part of the country. About updrafts. About downdrafts. About mist that formed in a flash, blinding you just as you were about to touch down . . .

The flying instructors were civilians. Or appeared to be. You called them *Mister* this, *Mister* that. Some of them looked ancient – and a few were, relatively speaking, with a lifetime of flying experience behind every move of stick or rudder. Many had learned their craft in World War I. At No. 8 EFTS, Vancouver, Rayne Schultz's excellent instructor, Harley Godwin, a druggist in peacetime, proudly wore his brass RFC cap badge pinned to his leather helmet. But many instructors looked little older than the students they had to train; in fact, some were younger. They were the "home grown" instructors, young men who had volunteered for the air force, dreaming of flying Spitfires and graduating with shining pilot's wings, only to find themselves at a flying instructors' school (FIS) sentenced to serve their country in Tiger Moths or Finches. Most objected vociferously, but they might have saved their breath. Once a student had been picked as instructor material, the powers that be seldom changed their minds about him – despite impassioned pleas and fervent assurances of total inability to demonstrate manoeuvres and go through the instructional "patter" at the same time. "One or two got away with this and wangled postings overseas," admits Group Captain Paul Davoud, assistant chief flying instructor at Central Flying School (CFS), Trenton, in the early days of the BCATP. "But we soon got wise to the situation and if someone didn't want to instruct we arranged a posting to a bombing and gunnery school as a target towing pilot and after that we had no trouble."[3]

The vast majority of those selected as elementary flying instructors swallowed their disappointment and, privately resolving to get overseas eventually, went off to learn the craft of teaching others how to fly. But more disappointments lay in store. As if it wasn't bad enough that fate had selected them to instruct, most of them soon found out that they were to become hybrids: half civilian, half serviceman, teaching air force students the basics of flying at *civilian* flying schools. Although officially still members of the RCAF, they had to revert to civilian status, on leave of absence and on a salary paid by the civilian school. They led curiously divided existences, being servicemen yet working in "mufti." Before long, a BCATP uniform appeared, a smart, dark-blue outfit with a peaked cap and the badge of the flying school company on the breast pocket of the tunic. Now instructors looked like airline pilots. Or meter readers. It all depended on your opinion of the uniform. The "Prairie Admiral" getup became the subject of much scorn – one account describing it quite unfairly as "a costume which, with individual embellishments, is worn by every able-bodied Pacifist in the Training Scheme . . . a source of amusement or embarrassment to the regular troops."[4] Initially, the part-time air force pilot-instructors never wore their air force uniforms, but from February 1942 on, they were permitted to wear RCAF blue *off* duty, an oddly appropriate reversal of the norm.

The youthful instructors got on with the job, trapped in work that few of them wanted and from which few would soon escape. The fact that they were a key element in making the all-important Allied air offensives possible provided little comfort, no matter how many times it was repeated; they hadn't joined the air force for *this*. "They were irked by the routine of elementary instruction," and "resented the prep-school campus atmosphere."[5]

Typical of the elementary flying instructors of that period was Herb Davidson. He had learned to fly before the war, at Leavens Brothers, on the now-defunct Barker Field in the northwest of Toronto, his home town. He obtained his private licence when he was nineteen. Two years later, in the summer of 1940, he was in the

RCAF, taking a course at CFS Trenton, which he completed on October 4. He rose in rank from AC2 to sergeant, but still hadn't been through a wings ceremony when he found himself at the airfield at Goderich, Ontario, known locally as Sky Harbour. He was wearing civilian clothes again, an employee of the combined Kitchener-Waterloo and County of Huron Flying Club, one of the RCAF instructors "on loan" from the air force. At that time the grass field had no hangars and no aircraft. With the first batch of pupils expected any day, the instructors hastened to Fort Erie to pick up factory-fresh Fleet Finches in which they would start training their pupils. Work began on October 16. Davidson was assigned five students, all of whom were initially billeted in the town, because accommodations had not yet been built at the airfield. He recalls that one student washed out, but the other four graduated and went on to SFTS. Three of the four eventually went overseas, distinguishing themselves on operations. "No two students were alike," he asserts. "You had to size up every one of them and act accordingly." Davidson was an elementary flying instructor for a year at Goderich, by which time he was becoming bored and keen to follow his former students on ops. He found that the authorities had no wish to let him go; he was too good at his job.

The best instructors were empathetic individuals who understood precisely the problems, real and imagined, that beset their students. Such instructors understood, too, that learning to fly takes some people longer than others and that the amount of dual instruction a person needs has little or no correlation with his potential. One of Davidson's students early in his instructing career was Dave Goldberg, who had heard about the declaration of war when on a Hamilton streetcar. Goldberg recalls that he was a "late bloomer" at EFTS. Fortunately he had an instructor of high calibre in Davidson, who was of the opinion that Goldberg had the makings of an excellent pilot, despite his less-than-speedy progress. Davidson was right. Goldberg graduated from EFTS, went on to service training, and became a highly regarded instructor and aerobatic pilot on Harvards at No. 6 SFTS, Dunnville, later going overseas as a Spitfire pilot with 416 Squadron RCAF.

On the face of it, teaching someone to fly an airplane does not seem to present any particular difficulties. The controls are logical enough, the stick moving backward or forward to send you up or down, from side to side – in conjunction with the rudder – to make you turn, and the engine in front tugging the aircraft along. The BCATP students had learned all about the theory of flight at ITS: how air streaming over the curved surface of a wing created lift and how the aircraft and its load created drag. It was all a matter of meeting one force with another, a great balancing act in the sky. Some compared learning to fly with learning to ride a bicycle. Certainly there were similarities, but the big difference was that the airplane travelled in an element that had absolutely no substance unless it was contorted at speed by the airplane's form. To fly meant to come to grips with the fact that you could never stop and take time to sort things out.

The instructor's first job was to introduce his pupils to their aircraft, to explain the functions of the controls, the ailerons, the elevators, the rudder, to point out how to check on the condition of the tires, the oil level, and the bracing wires, which seemed to be holding the whole frail-looking structure together. The Tiger Moths and Finches that predominated in the BCATP's early days were similar in size and appearance, their wings spanning slightly less than thirty feet; neat little biplanes with fixed landing gear and two cockpits enclosed by a sliding canopy to provide the occupants with a measure of protection from the elements. These diminutive trainers cruised at rather less than a hundred miles an hour and had a maximum speed of a little – *very* little – more. Painted bright yellow, as a warning to other traffic in the air and on the ground and to assist in locating the aircraft in the event of a forced landing, they were inevitably known as "Yellow Perils." A far cry from the Spitfires and Hurricanes that were the subjects of most students' ambitions, and a bit antiquated-looking to the discerning, the modest machines were perfect for the task at hand: that of introducing fledglings to the essentials of flight. Light, and possessing large areas of wing surface, they could be tricky to handle in windy weather – and, in fierce gales, occasionally became airborne while parked on the line. Like all good training aircraft, they were easy to fly but not that easy to fly well. They required constant attention, both in the air and on the ground.

De Havilland Canada of Toronto manufactured the Tiger Moth, a "Canadianized" version of the standard RAF trainer and one of a long line of Moth biplanes and monoplanes named by Geoffrey de Havilland, a noted lepidopterist as well as aircraft designer. The Finch was made by Fleet of Fort Erie, Ontario; students and instructors liked it but were less fond of its somewhat unreliable Kinner engine. Jim Ruddell of Georgetown, Ontario, did his elementary flying training on Finches at No. 11 EFTS, Cap de la Madeleine, Quebec, and recalls that "forced landings were not uncommon" due to engine troubles. Later the biplanes would be joined – and eventually replaced – by an attractive monoplane, the Cornell, designed by Fairchild in the United States and built under licence in Canada. The type, originally named the Freshman, had an unfortunate early career with the BCATP, with structural failures causing several fatal accidents. Investigation revealed that the Kaurit glue used to bond major components deteriorated in humid conditions. Slight modifications solved that problem, although many instructors considered the Cornell underpowered and *too* easy to fly and land. Another American design, the Stearman Model 75, was by far the biggest and most powerful of the elementary trainers used in the BCATP. The Stearmans saw only limited service, however; with no canopy over their cockpits, they were virtually useless in Canadian winters. At No. 32 EFTS, Bowden, Alberta, the RAF found that they could fly the Stearmans no longer than thirty minutes even in fall weather, despite the liberal use of grease on exposed portions of students and instructors.

Although the yellow trainers had looked small when first glimpsed from a comfortable distance, they seemed much larger close up. Larger but flimsier. Nudged by the breezes that seldom stop blowing over airfields, the aircraft creaked and groaned, an assemblage of struts and wires, with wings stretched taut and the smell of oil and fabric alive in your nostrils. An instructor – *your* instructor – leaned against the fuselage. It hardly looked strong enough to support his weight.

When instructor met student it was the beginning of a curious relationship, impersonal on the surface yet often possessing considerable emotional involvement as the two got to know one another and the intensity of the course began to take its toll. The routine seldom varied. The instructor walked around the aircraft, pupils in tow, pointing out the aircraft's features: wings ("Count 'em; there should be two, one on top, one at the bottom"), tail unit ("You'll find it fastened to the rear end"), the fuselage ("It's what the wings and tail are attached to and what you sit in"), and the engine ("That's supposed to be mounted up front, and if it isn't, complain. But not to me...").

Sitting in a classroom, you had experienced no trouble comprehending *why* the airplane flew, *why* the rudder deflected the air just as a ship's rudder deflects the water, *why* the elevator made the aircraft go up or down, *why* it was important that the ailerons be used to make turns smooth and skid-free. All quite logical; mere common sense, really. But suddenly you had a new and rather frightening perspective on the whole business. Lowering yourself uneasily into the aircraft, taking care not to push any knobs or catch your flying suit on any switches, you found yourself in a tiny seat in a form-fitting hole in the airplane's fuselage, your helmeted head sticking out, pimplelike. You peered thoughtfully at absurdly fragile-looking wings that seemed to stretch away to the horizon and at a nose that blocked all forward vision. Angled up at the sky, it appeared to be sniffing, taking the measure of the day. The peculiar airplane smell that had been a rather thrilling curiosity on the tarmac now enveloped you completely. Your world became redolent with the slightly sickening fumes of fuel and fabric, oil and sweat, and something indefinable that might have been the accumulated fears of all the students who had ever huddled in that seat . . .

The business of starting up the little Moths and Finches was an echo of pioneer days. No self-starters here. Long-suffering mechanics heaved on the two-bladed wooden propellers after exchanging bellowings about switches and ignition, raising thumbs, and waving hands. With a clatter the engine burst into noisy life. For the pupil about to embark on his first flight, the effect was remarkable. The aircraft felt as if it had suddenly developed St. Vitus's dance. Everything trembled: the seat in which you crouched – about as calmly as a prisoner in the electric chair awaiting the throwing of the switch – the joystick between your knees, the instrument panel before you, the tubular structure about you, the Perspex windshield and the transparent coupe top arcing over you. Communication between you and your instructor, a simple, civilized matter just moments ago, had abruptly declined into a confusion of burbles, some angry, some urgent, some impatient, but all totally incomprehensible. They emanated from something your instructor referred to as the Gosport, essentially a system of speaking and listening tubes connecting the two cockpits. The Gosport had appeared during World War I, the creation of an innovative if somewhat eccentric flying instructor, Robert Smith Barry of the RFC. Although the Gosport had its limitations, it was a considerable improvement over the shouting, gesticulating, and scribbling on scraps of paper that had been the means of communication between instructors and pupils up to that time. The Gosport system evoked images of a captain whistling down to the bowels of his vessel to talk to the engine room. Unfortunately, in the air the clatter of the motor and the screaming of the wind frequently made Gosport communications difficult, sometimes impossible. From time to time in chilly conditions, moisture in the tube would freeze, rendering the system useless; occasionally students threw up into their Gosport mouthpieces, with consequences that can be imagined.

The beginning of pilot training was for many students the most exciting time they had ever known. For months, years even, they had been in a fever of anticipation. They knew what to expect; they had read countless books and magazines and had endured every turgid moment of any film remotely connected with aviation. They had imagined the grass carpet streaming by on either side of the cockpit, hearing the screech of the engine at full throttle, feeling the thrill in the body of the aircraft itself, the trembling, the eagerness to be aloft, the clouds piled high like majestic castles shimmering in the heavens, exotic, exciting territories for you to explore . . .

Unhappily, first flights often had a way of squelching students' enthusiasm. All too many found flying far less enjoyable than they had anticipated. In fact they hated it. The din, the unfamiliar and thoroughly unsettling motion, the barrage of incomprehensible burblings in the earphones, the battering of the wind, the uncomfortable feeling that the frail little machine was about to disintegrate in mid-air, were bad enough. But they became far worse when the instructor took it into his head to indulge in steep turns and zoomings. The students felt their innards being alternately squeezed and squashed, their eyeballs threatening to vacate their sockets, their stomachs surging and plunging in direct opposition to the movements of the aircraft. Lord knows how many fledglings' desire to fly evaporated there and then.

When twenty-year-old Bill Swetman clambered into the front cockpit of a Fleet Finch at No. 3 EFTS, Crumlin Airport, near London, Ontario, it was a high point in his life, the realization of an ambition of long standing. The youngster from Kapuskasing in northern Ontario, with the clean-cut good looks of a blond Arrow collar man, had grown up dreaming of flying. Everything about aviation fascinated him. A substantial proportion of his spare time had been spent at the local airport, watching the infrequent comings and goings of aircraft, the Fairchilds and Fokkers and the sleek, silver airliners of the day. Swetman had had his future planned for years: a spell with the air force, learning to fly and accumulating hours and experience, then a career with one of the airlines at the controls of a big airliner, such as the Lockheed 14s that TCA had recently started flying into Kapuskasing. In 1938 he had attempted to join the RAF by enrolling in the "Trained in Canada" scheme. The RAF turned him down. Disappointed, he went to college in Montreal. He was there when the war broke out. Swetman lost little time in volunteering for aircrew, and now, in January 1941, he was about to step into an airplane and take the controls for the first time.

It was a pleasant day with little wind, ideal for an introductory flight. A mechanic swung the Finch's propeller and the Kinner B-5 radial engine caught at once, sending joyous little tremors through the biplane's airframe. Vibrating, pulsating, the Finch began to move. It felt as impatient to fly as the pupil occupying the front seat. Swetman revelled in everything: the snugness of the cockpit, the brushing of the propeller wash on his face, the gleaming instruments, the quivering needles, the sense of impatient anticipation in the vibration of the struts and wires. Then the supreme moment! Take-off! Its engine roaring, the biplane bumped across the grass field, gathering speed with every second. The Finch's tail lifted. The grass blurred – then fell away. Airborne! A magic moment!

Delighted, Swetman studied the movements of the controls as the Finch gained height. The ground turned beneath him like some meticulously constructed model, complete with minuscule train tracks and highways, dainty little trees, and lilliputian dwellings. This was unquestionably it!

Then he threw up.

The rest of the short flight was torture. Swetman tried to pay attention to the instructor's words booming and spluttering through the Gosport tube from the rear cockpit. But he found it impossible to think. The evidence of his disgrace was all over him and the cockpit. How had it happened? He hadn't felt sick. He hadn't had time. The air wasn't bumpy. In fact conditions could hardly have been better. So what was the bitter truth? That he was constitutionally incapable of being a pilot? It seemed so.

Back on the ground, Swetman had to clean up the mess in the Finch, as was the custom. ("You throw up; you wipe up.") That distasteful chore completed, he sat for an hour, sunk in gloomy introspection, contemplating the premature demise of his career in aviation. Perhaps he should quit there and then. Get it over with. But a fellow student talked him out of it. One accident didn't necessarily mean the end of everything. He wasn't going to give up so easily, was he? Swetman decided to try again. And to keep on trying, if they'd let him. On his next venture into the air, he took along a paper bag. Just in case. He didn't need it, though. Never again, in more than four thousand hours of flying, during an air force career that lasted a quarter of a century, did Swetman feel sick in the air. Today, he attributes the incident to *too* much enthusiasm, nervous over-reaction.

That first flight meant almost too much to him, as it did to countless BCATP students. Unhappily, many a keen recruit, no matter how splendidly he did in ground school, found that he couldn't cope with the realities of flying. The air force didn't have time to nurse borderline cases along. You had to learn and – even more important – *demonstrate* that you had learned. You had to show that you were able to co-ordinate the controls, maintaining a reasonably steady path through the sky – "straight and level flight," the instructors called it. At first you were convinced that it was impossible without four arms and an additional set of senses to keep correcting, pushing the stick one way, edging the rudder another, nullifying the effect of a gust of wind from one direction while compensating for a downdraft from another. Fledglings over-corrected. A mere nudge of the controls was all that was needed in those featherweight trainers, but the inexperienced hand did too much, sending the aircraft into a series of jagged changes of direction, bouncing it about the sky as if it were on the end of an elastic.

Students soon found that they performed best when they used the lightest possible touch on the stick. When they flew badly, they often discovered that if they relaxed their grip their flying improved at once. Len Morgan, an American volunteer from Louisville, Kentucky, who learned to fly at No. 9 EFTS, St. Catharines, Ontario, had a civilian instructor named Al Bennett, "a quiet, inoffensive fellow, anything but the shouting, abusive bully who too often found (and still finds) his way into this line of work." When, during their first flight together, Bennett handed over control to his pupil, Morgan grabbed the stick "and left my fingerprints there for all time." Morgan admits that his first turn at the controls was something less than a sparkling success; it was a significant moment nonetheless.

> We stumbled through the Ontario sky like a sick buzzard but I had control! Whatever crude path we cut through space was mine. It was me making that rattling old Kinner bob on the horizon, me pulling up that low wing, me slamming us both in our seats.... After just the right amount of this (he was a born instructor if ever one lived) I heard in my helmet, "Okay, I have control," and the ship settled instantly into normal flight. He flew us home and I marvelled at his skill as we dropped from the sky, shot across the fence at enormous speed and kissed the earth with an audible swish of rubber through grass. [6]

Alex McAlister approached his first flight in a state of near nervous collapse, with his head "almost bursting with excitement." But within moments of the aircraft starting up, a checkered flag was flying from the control tower. Flying had been scrubbed for the day. When at last McAlister took to the air on his first flight, the instructor lost no time in testing his reactions to aerobatics, diving and looping until the unfortunate pupil felt as if he were being telescoped.

> A piece of paper floated idly in front of my face, went up and settled on the coupe top, then slipped back to the floor where it belonged. I looked out and the ground loomed closer. Then all was serene. All, that is, except the waves of nausea creeping from beneath my belt to the vicinity of my throat. "The roll off the top. A combination of the last two maneuvers," announced Mr. Potts calmly. Control was impossible. The wave in my stomach lurched, surged forth, and hung in mid air along with the piece of paper. Nothing, not a thing, was affected by gravity. Mr. Potts viewed the scene in the mirror. "Lose something?" he asked, almost humanly. "Too bad. I guess this calls for an intermission."[7]

Student pilots found themselves in an unfamiliar world, where the instincts on which they had depended since birth had suddenly become a treacherous liability. The mother of legend who cautioned her airman son always to fly low and slow could be excused for thinking that she was giving him commonsense advice. On the ground, height and speed invariably spell danger. In the air they mean safety. Speed creates a kind of solidity in the air. Height is insurance. In order to gain height, you point the nose of your airplane up, but point it up too sharply and you will lose flying speed and you will plunge. To bring the aircraft into a landing, you operate the controls much as you would if you were starting to *climb*, the only difference being that you reduce your power. If your aircraft falls into a spin, you have to fight every instinct and make yourself point the nose *down* to get out of your stalled condition.

Students discovered that contradictions abounded in the air. They also discovered how persistently old habits kept intruding. If a landing approach seemed to be too fast, many a student pilot instinctively pressed on the rudder pedals as if they were the brake pedals in a car, just as, in an earlier era, student drivers going too fast were known to pull back on the steering wheel as if they were holding reins. The distinguished career of Marshal of the RAF John Salmond almost came to an untimely and messy end when he was learning to fly prior to World War I. In those days a first flight consisted of a "straight run, a hop, and down again," and as this was also before the days of dual control, it was not only Salmond's first time in the air but also his first solo. He had just one lesson on the operation of the controls, conducted in the safety of the hangar. Salmond successfully got the Anzani-Deperdussin off the turf and back again, but on the landing run he tried to stop the aircraft "by reining in, instinctively, as one does on a horse." To Salmond's dismay, his machine "reared its nose into the blue, stopped, dropped back to earth at an acute angle. . . ." Salmond was fortunate to emerge from the debris "with only a bruised head."[8]

In countless cases, a man's susceptibility to airsickness became a convenient deciding factor in sorting out the trainees. The borderliners with strong stomachs stayed in training, the others washed out. It was one way of deciding the matter, but an inefficient one. Bill Swetman was by no means the only pilot to experience queasiness and nausea when first subjected to loops and dives, or even straight and level flight; in fact, it is probably true that most trainees suffered similarly. However, nine out of ten eventually become inured to the peculiar motions of the air. Some cases were tougher than others. Marcel Beauchamp of Montreal was a student pilot at No. 11 EFTS, Cap de la Madeleine, Quebec, during the summer of 1942. He recalls that his instructor indulged in loops, stalls, and spins, then "dove down at a train. I felt sure he was going to hit it. He succeeded in scaring me and the engineer

down below, who was half out of the cab before we pulled up."
Beauchamp was violently ill. The authorities wanted to ground
him, but he insisted on being discharged from the RCAF if he
could not get into aircrew. He was sent to Trenton, and thence to
Toronto to undergo tests for airsickness. At that time, experi-
ments were being conducted with a tight-fitting inflated rubber
suit, the theory being that if an airman's blood could be made to
remain in his brain during violent manoeuvres, he would not
experience nausea. Garbed in this outfit, Beauchamp was "put on
a huge contraption in a hangar where some poor LAC had to
swing me for hours on end to see if they could make me sick." The
experimental suit was a singular success on this occasion, for in
spite of the best efforts of the perspiring LAC, Beauchamp failed
to produce the remains of his last meal – although, he admits, it
was a near thing. But he did not return to pilot training; instead he
remustered as a navigator, and airsickness was to plague him
throughout his air force career.

Later in the war, with a considerable surplus of aircrew building up in the system, there was little hope for pilot trainees with
weak stomachs. Instructors had orders to weed out the afflicted
without mercy. One student's progress in elementary flying training was slowed to such an extent by airsickness that he washed out.
But he got another chance, his instructor remarking, "Anyone
who had spilled as much of his guts . . . and then had some left
over, should be allowed to take the course again."[9] The student
was a young Englishman who spent any spare moment jotting
down stories and poetry – Arthur Hailey by name.

Students regarded the problem of airsickness with the sort of
dread usually reserved for fatal diseases. During flying training at
No. 7 EFTS, Windsor, John Macfie of Dunchurch, Ontario, con-
fided to his diary: "I hope it doesn't get me."[10] Not only students
were affected. An RAF trainee pilot, Ron Hunt, had an instructor
on Tiger Moths who was nervous, irritable, and "sometimes got
airsick."

One student pilot in four could be expected to fail EFTS. (RAF schools had a slightly lower rate of washouts, apparently due to the more elaborate method of pilot selection in Britain.) Some schools had a higher washout rate than others; it all depended upon the individual instructors – and upon the need for pilots. Although, early in the war, when the need for aircrew was urgent, many borderliners were given the benefit of the doubt and managed to pass through the system, by 1943 the pendulum had swung to the other extreme: the washout rate had risen to over 30 per cent at EFTSs. Now it was too high; the standards were being too rigidly enforced. The word from HQ was that "insufficient attention is sometimes given backward students and they have been too ruthlessly eliminated."[11]

It's an unfortunate truth that flying training brings out the worst in certain instructors. Absolute power corrupts absolutely, in the air just as on the ground. Students are totally dependent upon their instructors, not only for information but, in the early stages, for survival itself.

The harassed pupils of these airborne despots might have derived some comfort from knowing that the air force was far from satisfied with many of its elementary instructors. In an April 1941 report, Air Commodore Leckie, then director of training for the RCAF, declared in his no-nonsense way: "The need of the moment compelled us to produce an instructor for elementary training only. Such instructor is at a disadvantage in training pupils destined to fly Spitfires, Hurricanes, and other Service types, of which he, the instructor, has no knowledge and is quite unable to fly. The psychological effect is plain."[12] Leckie was also concerned by how monotonous an instructor's work was when he could only train students up to the elementary level: "He is at best a 'half-baked' product, his knowledge of aviation being restricted to Fleet or Tiger Moths, and the young pupil is quick to sense this limitation." Leckie thought that the most serious "disability" of all was that the civilian-trained instructor was expected to "inculcate a sense of discipline into his trainees when he himself, having been trained at a civilian club or company, has had no discipline." This lack of discipline, asserted Leckie, had been a major cause of the high number of "unnecessary accidents" in the previous few months. [13]

In 1941 the air force began – slowly, very slowly – to recall its elementary instructors "on loan" to civilian flying schools. The idea was to advance their training, commission suitable individuals, and give them the choice of returning to instructing duties at an EFTS or serving elsewhere, usually at an SFTS flying more advanced aircraft. It should have surprised no one at air force HQ that the flying schools loathed the idea. The manager of No. 7 EFTS, Windsor, Ontario, pointed out heatedly that the schools "should not be called upon to procure and train our own instructors and then have them raided by the RCAF. If such a system is to be inaugurated I can foresee nothing but disaster for the school. ... Our job here ... is to train pilots for active service, not to train instructors to become instructors at advanced school." [14]

Although the new policy undoubtedly caused problems for the schools, it was popular with the instructors and, despite complaints, it remained in place. The only objection of the air force instructors themselves was that it all took a long time to happen. Herb Davidson was one of the instructors involved. Selected by the air force for retraining, he had to leave Goderich and go to Lachine, Quebec, for a refresher course in air force drill and discipline, and thence to No. 6 SFTS, Dunnville, Ontario, for advanced training on Harvards, although he had specifically requested a multi-engine course. It was at Dunnville that he finally graduated – after instructing for two years. New Zealand student pilot Alan Gibson was at Dunnville when several Canadian sergeant instructors were learning to fly Harvards; some had up to 1,500 hours on Tiger Moths, he remembers.

Another proposal called for all civilian instructors to enlist in the RCAF; this stirred up even more vociferous opposition among the flying schools. The authorities quietly dropped the idea.

Every student had to go through the same sequence: take-off, straight and level flight, gentle turns, steeper turns, stalls and spins . . . then the tricky business of bringing the aircraft back to terra firma in one piece and the right way up. Problems with landing probably washed out more students than any other part of flying training. The landing process seemed simple. All you had to do was reduce the power progressively so that the airplane sank toward the field. Near the ground, you eased back on the stick as you reduced power. Although the aircraft wanted to settle, you held it off by raising the nose. Eventually, it could no longer fly; it settled, with barely a whisper as its wheels touched the ground.

So much for the theory. The practice was rarely so elegant. Students soon discovered that the air is seldom still; it stirs, it shifts, it burps, it sends out invisible fists to wallop aircraft levelling out to land, banging them into the ground with meticulously timed downdrafts, elbowing them off the threshold of the runway an instant before touchdown. No wonder many students regarded the air as implacably hostile, an element always lying in wait, eager to catch them off guard.

While landing problems were tough on the student, they could be even tougher on the instructor, for he was virtually a passenger. He had to sit there and watch it all happen. Most students misjudged their landings more than once, so that instead of levelling out just above the ground, they were too low and their wheels hit the ground hard, bouncing the long-suffering aircraft back into the air; or they levelled off too high and, when they ran out of flying speed, dropped, stonelike. The best instructors took no immediate action, but, demonstrating courage of a high order, waited to see how the pupil would cope, taking over the controls only if the situation became critical. RAF student "Cherry" Cherrington was impressed at how his instructor, Warrant Officer Wild, handled a dangerous incident at No. 34 EFTS, Assiniboia, Manitoba. On take-off in a Tiger Moth, Cherrington was startled to see another Tiger converging on him. "He was close enough to us," he says, "that when he broke away to his left, his wings hit ours, causing us to sideslip to the right. Our right lower wing tip touched the ground." Calmly, Wild proceeded with the take-off. Then, determined not to let his pupil's confidence be shaken by the near-disaster, he ordered Cherrington to handle the landing – "which turned out to be one of my better ones," Cherrington recalls.

Training aircraft needed sturdy undercarriages. Many a Finch or Tiger Moth came thumping back to earth from ten or twenty feet up. One of countless students whose landings tended to be adventurous was Jim Kelly of Toronto, a student pilot at No. 7 EFTS, Windsor, Ontario. During Kelly's enlistment medical, the MO had detected a depth-perception problem but had considered it curable. Kelly says, "I was instructed to (and did) perform daily eye exercises, consisting of holding a pencil vertically before my eyes and following it as I slowly moved it toward and away from my eyes." Unfortunately the treatment didn't effect a total cure. One day when Kelly was approaching for landing, he completely misjudged things. His Finch smacked into the ground, bounced high into the air, whirled into a dizzying ground loop, and ended up on its nose. Dazed, Kelly clambered from the wreck. He saw the crash truck roaring out to the scene, siren wailing. It squealed to a halt and the occupants leapt out, quickly checking that Kelly was all right and that there was no danger of fire. Then they appeared to lose interest in the incident. Without another glance in Kelly's direction, they clambered back into their truck and sped away. To Kelly's chagrin, he had to trudge back to the flight line, his parachute over his shoulder. It was his last flight as a student pilot.

Although Kelly was bitterly disappointed at being washed out, he recognized that "it may have saved a Lancaster and crew." (Later he remustered as a wireless operator/air gunner and was a member of the crew of 419 Squadron's Lancaster A-Able, in which Andy Mynarski won the Victoria Cross for his heroism on the night of June 12/13, 1944, when he attempted to save the life of the rear gunner trapped in his turret.)

The aim of the elementary flying training program was to produce a pilot with the basic skills, a man who felt at home in the air, who no longer had to *think* about every movement of the controls, who had in fact begun to fly instinctively, easing stick and rudder pedals to counteract a downdraft on final without having to reason it all out, an airman who could get himself out of stalls and spins and who knew what to do (and what not to do) when his engine failed and he had to find a place to put down. A few students were "naturals" who took to every aspect of flying as if born to it; the majority had to work hard, sitting for hours in parked aircraft, rehearsing the actions they would take in this situation or that. Chances are the plodders were safer pilots, because they had so much respect for the air and its dangers.

Confidence also played a major role in the development of flying skills, and a sudden loss of confidence could have dire consequences. By the fall of 1941, Tet Walston of Morley, Yorkshire, had finally realized his ambition of joining the air force and learning to fly. Full of anticipation, he arrived at No. 31 EFTS, then based at Calgary. (It was, he recalls, a large grass field, "and TCA flew Lockheed 10s in and out. *They* had radio but we did not. The mind now boggles at the idea of inexperienced u/t [under training] pilots being mixed up with the big boys, but we all survived.") Walston enjoyed exceptional progress. He was ready to solo at five hours, but another student with about the same number of hours crashed on his first solo. The chief flying instructor (CFI) immediately increased the minimum number of hours required before solo. Although Walston's instructor, Sergeant Hunter, had intended to send his pupil solo that same day, now the great occasion had to be put off. This had a strangely degenerative effect on Walston's flying. He experienced a phenomenon not unfamiliar in flying training: above-average progress followed by an abrupt loss of confidence and an immediate decline in flying skills. About this time, the school moved to a new location at De Winton, a recently completed airfield south of Calgary. Walston had slipped behind most of his class, his take-offs and landings getting worse by the day. "Time was running out for me. Ten hours was scrub time, and still I hadn't soloed." The situation looked bad. And it got worse. His instructor, Sergeant Hunter, was posted away. Fortunately for Walston, Hunter's replacement, Sergeant Lithgoe, was as empathetic as his predecessor; he was able to bring Walston out of his slump, building up his confidence, praising his steep turns, stalls, and spin recoveries, working tirelessly on take-offs and landings. "I finally soloed at 12 hours, 15 minutes," says Walston. "Whew, that was close!"

A pilot's first solo is a milestone in his career. None ever forgets the magic moment of being aloft, in complete control of the aircraft for the first time. The instructor, formerly so omnipresent, is now a tiny, forlorn, and rather lonely figure far below – a mere midget, standing, watching, worrying as all mother hens are supposed to worry when their chicks leave the nest for the first time.

Bud Hudson from Falls Church, Virginia, had a demanding instructor at No. 7 EFTS, Windsor, who "felt the need of giving me hell from takeoff to landing." When Hudson took off on his first solo, he revelled in being in complete command – but he swears he could hear his instructor's voice, still criticizing, still complaining. Len Morgan, another American volunteer at the same school, turned in his seat twice during his first solo to make sure his instructor really was absent, and he "laughed loud enough to hear myself over the Kinner's roar. When I walked into the hangar a few minutes later, Mr. Bennett [Morgan's instructor] was trying hard to look relaxed. Someone asked him if he usually smoked two cigarettes at once." [15]

Mr. Pearson, Murray Peden's instructor at No. 5 EFTS, High River, Alberta, made sure Peden was mentally ready before sending him off solo. After two circuits, Peden approached for his third landing, but as he levelled out, his instructor jerked the stick forward, bouncing the Tiger Moth on the ground, sending it up into the air. Peden took control and made it safely back to earth. Pearson ordered another circuit. On his next approach, Peden had the aircraft lined up for a good landing when Pearson pulled the stick back. The Tiger soared. Peden had to respond quickly to prevent a stall. After that, he landed and "taxied back to the control box in that mindless looking zig-zag, like an earthbound dragonfly run amok, and again I turned out of wind at Pearson's hand signal." The instructor clambered out of the aircraft and told Peden to do a circuit solo. The great moment had finally arrived! A few minutes later Peden brought the Tiger in for a perfect landing: "I glowed with contentment. My log book would now show nine hours and 35 minutes dual and *15 minutes solo*." [16]

No student pilot took the occasion lightly; neither did his instructor, for the flight was as tough on the instructor's nervous system as on that of the student. Most instructors sent their students off solo after eight to ten hours of dual flying. A few quick learners went solo after six or seven hours; a few more didn't go solo until eleven or even twelve hours. Some had still more time. Rayne Schultz, who had turned down a hundred dollars to give up his place at EFTS, had "two left feet," he says, when he started flying, and although he later became one of the outstanding night fighter pilots of the RCAF, it took nearly eleven hours of dual instruction before his ex-RFC instructor, Harley Godwin, sent him solo. Chuck Appleton of Cambridge, England, had over fifteen hours of dual instruction before going solo. Clearly, the number of hours a student had before soloing had little to do with ability, for Appleton proved to be good enough to be selected as a flying instructor; later he flew a tour of ops with Bomber Command, winning the DFC.

Few would dispute Herb Davidson's assertion that all pupils were different. But there may have been two notable exceptions: the identical twin Warren brothers, Bruce and Douglas, from Nanton, Alberta. When they arrived at No. 5 EFTS, High River, in July 1941 to commence flying training, no one – including their flying instructor – could tell them apart. The Warrens soon became known as Duke Mark I and Duke Mark II. Interestingly, not only did the two Dukes look identical, there was also nothing to choose between them as far as their flying ability was concerned. Throughout the training process, they made progress at exactly the same rate. Inseparable, they shared everything – including clothes – studied together, relaxed together. Sometimes they substituted for one another in training exercises. Curiously, this led to one of the few instances in which their progress differed a little. Douglas (Duke Mark II) recalls that he usually got a B in Link training "as himself," but if he masqueraded as his brother, he invariably got a C! Throughout their flying training, their instructor was never quite sure which Duke he had in the aircraft – which was precisely the way the twins wanted it.

Jim Emmerson of Georgetown, Ontario, learned to fly at No. 4 EFTS, Windsor Mills, Quebec. He recalls:

> Anyone who had not flown solo after ten hours of dual instruction was likely to hear the dreaded words of doom: "You'll have to go up with the CFI." A flight with the chief flying instructor was invariably the formality preceding washout and remustering to some other aircrew trade. My instructor, Pilot Officer J. C. Stairs, promised me at the ten-hour level that if I made a good landing that day I could go solo. Could I make a good landing? When I levelled out, the Tiger Moth

dropped as if filled with cement. With it dropped my
hopes of my first solo. Stairs uttered the final and fatal
sentence about the CFI and we climbed out and headed
back to the flight shack. My face was so long that had it
been pictured on page one of a newspaper, the editor
would have been compelled to continue it on page two.
At the flight shack, Stairs turned to me and demanded,
"Do you really think you can take up that thing alone?" I
was as eager as a puppy freed of its leash. "I'm positive I
can do it, sir," I exclaimed. "OK, off you go, then," said
Stairs as he disappeared into the flight shack. The plane
had to be refuelled before the flight, which now seemed
to me as important as Lindbergh's flight over the Atlantic. In gassing up, the mechanic accidentally spilled a
considerable amount of fuel over the wing and engine
cowling. I was not daunted. I was prepared to go down in
flames before I'd give up on this attempt. After mopping
up, I trundled out, headed into wind, pushed open the
throttle and roared aloft. I got back to earth with only
one modest bounce. June 17, 1943 – hardly a day that will
live in history, but a day noted in big, red letters in my
logbook.

No matter how well a student might have performed in dual
instruction, there was no telling how he would react to being
alone in the air for the first time. Len Morgan recalls one student
who couldn't land. "Thirty or more times he made the circuit,
executed a beautiful approach and then crammed everything to
the firewall just before touchdown. They grounded the rest of us
while an instructor flew alongside him trying to lead him to earth.
It didn't work. Finally, just at dusk, the frightened boy dropped
onto the grass in a perfect three-point. The next day he was on his
way to gunner's school." [17]

Morgan saw a British student stall his aircraft about thirty feet
above the ground.

The Fleet hit on all three wheels. . . . The fuselage bent in the middle, the wings sagged, the gear crumpled. Out of the cloud of dust the Fleet staggered on flat tires, wires trailing under the broken tail, smashed prop beating the ground. The student taxied to the line, swung around and parked neatly between two intact trainers, stiff upper lip to the end. He became a navigator. . . . Another fellow did a fine job until his solo. He circled the field once, landed nicely, taxied in, shut down and went to see his CO. "Sir, I don't want to be a pilot," he said . . . [18]

Another American volunteer, Nick Knilans, took his training at No. 13 EFTS, St. Eugene, Ontario. He remembers his first solo vividly, for he landed his Finch "in a strong crosswind, tipped up onto one wing and over onto my back." He got another chance.

Two students at No. 12 EFTS, Goderich, Murray Marshall and Andy Carswell, set off simultaneously on their first solos. Concentrating on the task at hand, neither noticed that their take-off runs over the grass field were converging at an alarming rate. Only at the last moment did either student notice the presence of the other. They missed a collision by an instant.

All went well on Cliff Stead's first solo at No. 20 EFTS, Oshawa, Ontario, until he came in to land. "After flattening out, I still had too much flying speed, and as a result, the aircraft began to climb. As I failed to correct . . . the aircraft went into a steep nose-up attitude and stalled." In the ensuing crash, the Tiger Moth's top wing collapsed. Stead himself was lucky enough to emerge unhurt. He was also lucky that the base engineering officer had witnessed the incident and declared that the aircraft's struts should have withstood the shock. Stead was allowed to try again.

Earl Roberts took off on his first solo and followed another Tiger Moth in the circuit. He hit the other aircraft's wake turbulence, which caused a sudden dropping of one wing. "In my agitated state," says Roberts, "I got the controls crossed, with left aileron and right rudder. . . . Instead of turning left . . . the aircraft did a 360-degree turn to the right. . . . If it hadn't been for the inherent stability of the Tiger Moth, I would likely have flicked into a spin." Roberts landed, expecting to be washed out. Instead, his instructor – "whose face was as white as mine" – sent him off on a second solo, which he successfully completed.

At No. 15 EFTS, Regina, local legend has it that one student departed on his first solo in fine form but couldn't face the prospect of landing. After several circuits, he took the Tiger Moth to three thousand feet, clambered out of his cockpit, and jumped, preferring a descent by parachute to trying to land the plane.

When Alex McAlister was told to go solo, his hands began to tremble uncontrollably. Sweat poured from his brow yet his hands were like ice. When he took off – "hanging on to the stick with the grip of a drowning man" – he quickly settled down and began to enjoy the flying, feeling "like a bird who had escaped from a cage." [19] The tension of the great occasion had odd effects on individuals. Fleet Air Arm student pilot Bill Martin seemed to be fully conscious of taking off on his first solo, but the rest of the flight didn't register, including the landing. "I must have gone through all the correct motions somehow . . . but I can't remember a thing about it now," [20] he writes. (A curiously similar incident occurred to another Fleet Air Arm pilot, Jim Smythe, also a Goderich graduate. Flying a Wildcat fighter at the fighter school at Yeovilton, England, he was dazed when the plane's hood broke loose and knocked him on the head. "I have no knowledge of landing the plane," he comments. Habit came to his rescue at a dangerous moment; automatically, he carried out all his checks and brought the Wildcat in for a perfect landing.)

Robert Brady had no trouble completing his solo at No. 17 EFTS, Stanley, Nova Scotia, but shortly after the great event, he ran into a classic problem. The engine of his Finch failed while he was doing circuits and bumps.

Just like in the old flying movies, the cockpit filled with smoke, the rpm dropped off and the engine made unpleasant noises. Before panic set in I automatically set about doing the right things, I hoped ... shut off the ignition and went into a straight glide ahead. .... My day was not improved when I saw several stone walls breaking up what I had thought to be a perfect landing area. Mind-changing was a no-no when in a powerless glide, but I had just enough height and airspeed to make a very gentle bank to starboard and head for a large cultivated area. This approach took me rather close to the chimney of a farmhouse but lined me up with a nice, large, soft ploughed area. As I was holding off, my undercarriage cut through a wire and post fence, but did not spoil a neat three-point landing. The Fleet rolled about ten feet in the soft earth, which turned out to be a crop of cabbages. The silence was beautiful, broken only by the birds and the beating of my heart. I felt terrific. Peace was short-lived. It was shattered by three children who I gather had been sitting on the wire fence about fifty feet from where I cut through. They had been deposited in a heap and were shouting in great excitement. Then their father approached my plane calmly puffing on his pipe and said quietly: "Who's paying for my cabbages, son?"[21]

At some EFTSs custom called for students who had soloed to be tossed into a water tank – or the shower, depending on the season. At others, the favoured few sported polka-dot scarves with their flying togs. The next big step in the training process was the much-feared twenty-hour check with a senior instructor. Those who successfully overcame this hurdle could be found wearing Tone-Ray aviation-style sunglasses even on the dullest days. But the silk scarves and sunglasses were only interim badges of accomplishment. Every student pilot ached for the day when he could sew on his pilot's wings, that shapely symbol of accomplishment that set you apart from the rest of mankind for ever.

Students had little time to dwell on the triumph of their first solo flights. As far as the air force was concerned, they were still *sprogs,* mere fledglings possessing only the most basic of skills; they had a long way to go before they were truly professional airmen. Instructors seemed to be tougher, more critical now. And not without reason; instructors knew all too well that many students tend to become a little overconfident after their first solos, imbued with a feeling that they have single-handedly conquered the air and all its mysteries. Another danger lay in the natural inclination of young men to attempt to impress one another. ("Flick turns? Sure, nothing to them. I've been doing them for *weeks.*") RAF student Len Dunn, at No. 35 EFTS, Neepawa, Manitoba, heard advanced students talking about the wonders of aerobatics and couldn't wait to try his hand.

> In solitary state I climbed the Tiger steadily to four thousand feet, dived to increase speed, pulled up into the loop, and replaced earth with sky. My seat belt was too loose; gravity asserted itself, I dropped clear of the rudder pedals, and, deluged with dirt from the cockpit floor, lost control. The earth was revolving around my head . . . I had stalled into an inverted spin. Instinctive reaction came to my rescue, I stopped the spin and half-looped back to normal flight but far too close to the ground for comfort!

Students developed their own tests of aeronautical maturity. At No. 20 EFTS, Oshawa, Ontario, two prominent trees stood near the field, little more than a Tiger Moth's wing span apart. No student could call himself a real pilot until he had flown between them. At No. 13 EFTS, St. Eugene, Jeff Mellon recalls that the initiation into the ranks of the stars was to do circuits and bumps on the nearby Montreal-to-Ottawa highway in the early-morning hours when no traffic was in the immediate vicinity.

The Finches and Tiger Moths (often referred to as "Tiger-schmitts") at EFTS bore a striking resemblance to the aerial mounts of the aces of the Great War. No one should have been surprised, therefore, that so many students had the idea of staging mock dogfights with their friends. The game was to find an unoccupied corner of sky, somewhere safely out of sight of anyone, then, with dives, banks, and rolls, see who could get on the other's tail first. Nine out of ten students seem to have had the same notion. The authorities discouraged such ambitions, only too aware of the dangers of inexperienced pilots cavorting about the sky. While at No. 9 EFTS, St. Catharines, Ontario, Allan Caine and another student decided to indulge in an aerial duel over a racetrack near Fort Erie on the Canadian/U.S. border: "We felt comparatively safe, as we were well away from our base." But during the mock dogfight, Caine's friend dived on what he thought was Caine's Tiger Moth. "He suddenly realized that there were two occupants in the airplane and that one of them had to be an instructor," says Caine. The two students quickly scattered. But it was too late, the dogfight had been witnessed. "When we returned to base," Caine recalls, "my friend was confronted by the instructor." He demanded to know the identity of the second pilot. The student refused to identify Caine in spite of threats of being washed out. The two students had been fortunate indeed. They hadn't killed themselves playing Billy Bishop and von Richthofen and, although an instructor had discovered them in their illegal pursuit, he decided not to report them to higher authority.

The sheer delight of being in complete command of an aircraft spurred young students to other imprudent acts. At No. 13 EFTS, St. Eugene, Don McTaggart of Toronto was enjoying a little solo flying when the cockpit cover of his Finch slid back and jammed. A frigid wind battered McTaggart, so he decided to change seats.

Unbuckling his harness, he pulled himself out of the front cockpit. Still grasping the stick with one hand, he draped himself over
the narrow coaming between the cockpits while he groped for the
instructor's stick with the other hand. The transfer was proving
trickier than McTaggart had anticipated. He became uncomfort-
ably aware of the fact that he was five thousand feet above the
earth. Then he became aware of something else: another Finch
had chosen that inconvenient moment to appear on the scene –
and it contained two heads, one of them unquestionably belonging to an instructor. McTaggart lost no time in wriggling back into
the front cockpit and resuming his flight. Whatever the instructor
thought of McTaggart's antics, he took the matter no further. Not
so fortunate was Clyde East, a volunteer from Chatham, Virginia.
At No. 4 EFTS, Windsor Mills, Quebec, an instructor spotted him
indulging in aerial horseplay with another student. East was on
his way to Trenton the next day.

Like casualties on an operational squadron, the "washouts"
vanished one by one. There in the morning; gone in the afternoon. No farewell parties sent them singing on their way;
everyone, victim and survivors, seemed to want to get the nasty
business over and done with as quickly and painlessly as possible.
Besides, it was hard to escape the uneasy feeling that the washout's
problem might be contagious. Although he may have been a
prince of a guy, it was maybe better not to have too much contact
with him in these last few hours . . .

On the other hand, many students were far too busy with their
own training to notice what was happening to their fellows. Bill
Martin, former Fleet Air Arm student pilot, recalls: "Out of our
original course of forty, fifteen failed. . . . It is an incredible fact,
but I was totally unaware of their exits at the time." [22]

For most students, life after the first solo was circuits and bumps,
hour after hour of them, sometimes with an instructor, more

often alone. They learned about forced landings, cross-wind landings and take-offs, basic aerobatics, the fundamentals of instrument flying, and other mysteries – and at No. 7 EFTS, Windsor, that included the design features of the aircraft they flew. Bud Hudson recalls a student there who was flying solo in a Finch when he accidentally jettisoned his canopy. Glancing back, the student saw that the canopy had fallen free of the aircraft but that on the way it seemed to have damaged the vertical fin and rudder, knocking the assemblage off-centre. Alarmed, he bailed out. Later he was informed that this offset was built into the Finch's design to counteract the torque generated by the propeller. He had abandoned a perfectly good aircraft. Hudson remarks that the incident prompted a "substantial increase in emphasis on the theory of flight and aerodynamics" in ground-school lectures.

Sometimes students found it hard to find time for flying and ground training. Fraser Gardner, a student pilot from St. Catharines, Ontario, was practising spins at No. 9 EFTS, in his home town, "when I realized that I was going to be late for ground school." The realization came when he was at an altitude of six thousand feet. Rapidly he spun down to one thousand feet, landed, and ran off to ground school. But to his horror he found he was stone deaf. He was ordered to take off immediately, climb to six thousand feet again, and come down slowly. "This relieved the air pressure and I could hear again."

Although much ground training was valuable, too much time seemed to be spent on subjects of little use to pilots. Armament, for instance. "I thought it was a waste," comments former BCATP student and instructor Bob McBey, a Scot from Ardersier, near Inverness. "I could never understand why I had to be able to strip a Browning machine gun, clear a stoppage, etc., when the opportunity for a pilot to do so in action was pretty well *nil*."

Soon after first solo came the time to learn something of the mysteries of flying on instruments – *blind* flying, the air force called it. The Tiger Moths and Finches had a form of hood that would enclose the student's cockpit; the rear in the Moth, the front in the Finch. With the blind-flying hood in position, the student was in pitch darkness, no longer able to see the familiar horizon which for so long had obligingly informed him that the sky was up there and the earth down there. Now he had to refer to a collection of instruments for that vital morsel of information, as well as for such important facts as his height, his speed, his direction, and his attitude in relation to Mother Earth. Enveloped in noisy, vibrating darkness, the student found out how unreliable his basic senses could be.

In fact, they could be downright dangerous. The reason was a condition called "vertigo." BCATP instructors drummed into their students that their lives depended upon their instruments. They had to be believed, utterly, without question. The instructors were right to lay such stress on the point, but few explained why. It was all very well for students to nod and agree never to question the veracity of their instruments, but, given sufficiently stressful circumstances, a man's senses are quite capable of convincing him that they are telling the truth and all those little mechanical contrivances are lying. Or misinformed. Or just plain broken. In normal flight, a pilot makes many tiny corrections and never thinks about them. In daylight they are of little more significance than the automatic adjustments of a driver on a city street. But in darkness the whole process can be infinitely more dangerous. Turning, a pilot can be fooled by his senses into believing that he is diving; in a turn, centrifugal force presses the body into the seat, as it does when an aircraft pulls out of a steep descent or climb. Thus, a pilot in darkness might think he is ascending when he is in fact simply banking. Returning to straight and level flight produces a reduction of pressure on the seat, giving the pilot the distinct impression that he is starting to descend. Vertigo has a multiplicity of such ways of bamboozling the body's senses, most of them potentially fatal under certain conditions.

Jim Emmerson recalls "intense mental pressure" to obey his instincts, which might tell him that he was tipped at an angle or going around in a turn when he was flying straight and level. "I had to fight against panic and the fear that I was losing control of my senses, which could lead to losing control of the aircraft. Fortunately, things never went that far with me, but I sometimes wonder how many crews may have perished because a pilot gave in to panic under such circumstances and lost control. . . . Had someone during my training warned me about vertigo, explained what it was and how it was caused and how to cope with it, it probably would have made my lot somewhat easier."

It might in fact have been wise to explain to students that their bodies were very good at detecting motion on solid ground, but far less efficient in the air. A person's equilibrium depends upon three tiny, semi-circular tubes in each inner ear. Lined with fine hairs and filled with liquid, they are the body's spirit levels. They provide accurate reports under normal circumstances, in which relatively slow movements are involved, where turns are ninety degrees or less, and where the surface – that is, the earth – is stable. All too often in flight, those conditions are not met. Hence the need for blind-flying instruments.

For BCATP students, learning to fly on instruments was like learning to fly all over again. Instead of gazing ahead at a sunlit horizon, you had to keep your eyes glued to impassive dials that dutifully recorded your attitude in the sky no matter whether you were proceeding in splendid style or were about to hit the ground in an inverted spin. If you could keep the needles of those instruments under control, you could be reasonably sure that you were the right way up and flying straight and level. It was when you essayed a turn that things became complicated. Eyes never straying from those quivering needles, you eased the controls into gentle left or right turns as you tried to picture what was happening to the aircraft in relation to the fondly remembered horizon. Many students found it as challenging as learning to juggle. Alex McAlister was one. The first time he tried his hand at instrument flying, he nearly stalled the aircraft, to the intense displeasure of his instructor, Mr. Potts.

"'Look at the god-damn airspeed. What the hell do you think this is – a tractor? You're bending the aircraft into a horseshoe – get the needle into the centre – the ball is way out. Oh, my gawd, we're going into a screaming spiral dive! Don't be surprised if the engine falls out. Look out, you're way past the maximum rpm. Push back the hood. I have control,' said Mr. Potts weakly." 23

Some pilots reacted unexpectedly when robbed of the sight of the familiar horizon. Former BCATP student pilot Len Dunn recalls a pupil and instructor at No. 35 EFTS, Neepawa, who took off "blind." Something went wrong during the take-off run, however, and the Tiger Moth veered off track. It flew "straight through a hangar window, shed its wings en route," and came to a halt lodged in the sturdy roof beams high up inside the hangar. Although miraculously unharmed, the two occupants had to sit motionless for some time, not daring to move until help arrived, for fear of disturbing the wreckage and sending it crashing to the floor forty feet below.

Few students left EFTS without a class photograph. Jim Emmerson had *two*. He recalls that fully 50 per cent of the students had been washed out during the course. The surviving students arranged themselves in front of one of No. 4 EFTS's Tiger Moths, in two groups. There was plenty of room for the entire class before a Tiger but, says Emmerson, "an overzealous administrator decreed that it would be a security breach to have our entire graduating class taken in one photo." Two of the students helped to confound German Intelligence by inserting themselves into both shots.

Billy Bishop spent a good deal of the summer of 1940 in London. He was a popular figure in the British capital. He had his own table at Claridge's and was a frequent guest at the prime minister's country house, Chequers. He wanted Churchill to find him an active role in the war, but the PM was of the opinion that Bishop's greatest contribution to the war effort would be to encourage recruiting for the BCATP. At the time, the news from the Mother Country was dominated by the Battle of Britain. Although many later aerial conflicts surpassed it in numbers of aircraft and casualties, few would equal its strategic significance. The battle changed history, because it prevented the Nazis from invading Britain, thus preserving the base from which Nazi Germany was eventually defeated. Had Hitler occupied Britain, it is hard to imagine the course the war would have taken.

Canadians saw the Battle of Britain in the same way as did the British public: the gallant, outnumbered RAF cockily shooting down everything the hitherto unbeatable *Luftwaffe* sent their way. A poke in the eye for the overweight, over-medalled Göring. A blow to Adolf's pride. It all smacked satisfactorily of Drake and the singeing of the Spanish king's beard in Cadiz Harbour. Clearly the Hurricanes and Spitfires of Fighter Command and their pilots were more than a match for the swaggering Germans. The numbers proved it. . . . In fact, the numbers proved nothing, for they were grossly inflated, partly because of errors and duplications by those involved in the daily battles, partly because the British Ministry of Information wanted it that way. It was good for morale. And, whatever the numbers said, the battle had been won. By mid-September the German navy stopped assembling transport vessels and began to disperse them. Operation Sea Lion was shelved.

The *Luftwaffe* turned to night bombing. For the first few months of the war, the British and the Germans had refrained from bombing each other's cities. Both sides held back from that final step into total war, perhaps believing that a negotiated peace was still possible. But during the summer of 1940, a German bomber dropped bombs on London, almost certainly by accident. Churchill lost no time in ordering immediate attacks on Berlin. The strategic bombing campaign had begun.

In those early days, the RAF's bomber crews flew by moon-light, the captain of each Whitley, Hampden, or Wellington largely responsible for deciding bombing heights and flight paths. The era of the massive bomber streams, in which as many as a thousand aircraft flew on precisely the same course and at the same altitude, was a couple of years away.

On a bitterly cold night in November 1940, a twin-engined Whitley bomber of 102 Squadron approached Cologne. The crew could see the Rhine glinting below like a shimmering serpent in the bright moonlight. The Whitley's intercom had failed en route to the target, and the second pilot had been obliged to act as run-ner, hurrying back and forth along the long fuselage, carrying messages from the skipper to the navigator and other members of the crew. At last the aircraft settled on its bomb run. The crew con-centrated on the job of delivering the bomb load on target. It was almost time to drop them . . . just a few seconds more. The moments ticked away as the bomber traversed the city. The observer's thumb began to press the release button when, sud-denly, shockingly, flak found them. A shell smashed into the air-craft's front turret, the explosion wrenching the control column from the pilot's hands. The bomber plummeted, acrid, choking fumes filling the cockpit. Struggling to see his instruments, the skipper fought to regain control. A skilled and experienced pilot, he managed to pull the heavy bomber out of its dive. But no sooner had he levelled out than a second shell exploded just above the port wing, sending red-hot shards of shrapnel slashing through the side of the fuselage and hitting a rack of marker flares. With a blinding flash the flares exploded, showering flame over the wireless operator crouched beside the flare chute. The Whitley staggered, fire roaring from its ripped fuselage, like some mon-strous Bunsen burner in the sky. Again the skipper fought to regain control. Luckily, the icy wind that burst through the torn fuselage helped to extinguish the flames. The Whitley limped away from the flak and the searchlights.

Then the skipper discovered that the aircraft's bombs still nestled in their racks. In the excitement they had been momentarily forgotten. Without hesitation, he turned back over the city, lined up for another bomb run, and had the satisfaction of seeing the load fall on the Cologne marshalling yards. At last he turned for home. The aircraft sagged and flexed, apparently about to come to pieces in midair. But it held. At daybreak the Whitley landed at 102's Yorkshire base, Linton-on-Ouse. The skipper was a cheerful, supremely self-confident young man named Leonard Cheshire. His rear gunner, a tall, pipe-smoking Cornishman, Richard Rivaz by name, also came through unscathed. But, after two tours of ops sitting in the tail turret, he decided it was time for a change in his air force career. He applied to remuster as a pilot. Accepted, he was soon on his way to Canada to learn to fly with the BCATP.

CHAPTER FOUR

Advanced Flying Training

North American Harvard II

"The word 'danger' should never be used in training aviators. Fear must be overcome, for it is the deadliest foe to knowledge. Nothing a student may do in the air is dangerous, if he knows what he is doing and what the result will be. Ignorance is the cause of most accidents . . ."
    – Robert Smith Barry[1]

By the time you completed elementary flying training, your logbook had begun to lose its *ab initio* freshness. It would take many more months and scores of flights to acquire any sort of long-service patina, but the process had begun. Now your aerial experiences filled several of its pages, every one neatly listing date, aircraft type and number, names of occupants, "duty" accomplished, with the time of each flight and the accumulated totals of dual and solo time all recorded and periodically verified by your flight commander. It added up to sixty to seventy hours in the air.

In World War I, pilots went off to fly in combat with fewer hours under their belts. But times – and airplanes – had changed. Once a BCATP student had mastered low-powered elementary trainers, he had only begun the learning process. Now it was time to progress to more challenging aircraft, high powered machines with variable-pitch propellers, retractable undercarriages, and other technical advances. Whereas a mere half-dozen instruments had stared unblinkingly at you from the panel of a Tiger Moth, now you were faced with half a hundred. Students newly arrived at service flying training school (SFTS) spent hours in the cockpits of their aircraft, studying every dial, every lever and switch, striving to become as familiar with the layout of these big, powerful craft as they were of the far simpler Moths and Finches. It seemed impossible.

When Len Morgan arrived at No. 14 SFTS, Aylmer, Ontario, he encountered the Harvard. It was "a huge, all-metal creation with a tremendous 600 horsepower Pratt and Whitney nine-cylinder radial glistening darkly under an enormous cowl. . . . There was a feeling of brute strength about it." [2] New Zealander Alan Gibson thought the Harvards were "aggressive-looking beasts" when he first saw them at No. 6 SFTS, Dunnville, Ontario. Fellow Dunnville student Frank Phripp of Toronto was similarly intimidated, describing the Harvard as a "frightening psychological jump" from the Tiger Moths he had been flying at EFTS. It was an aircraft that "demanded respect and would bite back if its aerodynamics were not catered to properly."

The Yale looked like a Harvard with a fixed gear, but in fact it was a considerably less powerful aircraft, and far less popular. It had been ordered in quantity by the French air force, and after the collapse of France some Yales found their way to Canada, where the RCAF used them as intermediate trainers to prepare students for the more demanding Harvard. Soon it became obvious that the intermediate stage was unnecessary and wasted training time. Thereafter, most of the Yales found employment on radio training duties.

A Canadian design known as the Fleet Fort had been developed to meet the need for an intermediate and advanced trainer, but it was not a success. While serving as an instructor at No. 2 Wireless School, Shepard, Alberta, Jack Merryfield investigated fourteen forced landings by Fleet Forts in a period of little more than three months, "usually from defective oil lines, leaking oil tanks, or an overheated engine, etc." During a solo flight near Shepard, Merryfield observed the fuel-tank cap on the starboard wing of his Fort slowly unscrewing because of the vibration. Acutely conscious of the fact that the exhaust pipe was also on the starboard side and that if the cap came off fuel might be sucked out in its general direction, he "lost no time returning to Shepard . . ."[3] He landed safely. Surviving Forts were soon replaced by Yales and Harvards.

Most of the twin-engined trainers used in the BCATP were British-designed Ansons, roomy and good-natured, but still a big step up for pilots accustomed to the narrow confines and basic instruments of a Moth or Finch. Walter Miller of London, Ontario, experienced "something of a shock" when he first saw the instrument panel of the twin-engined Anson aircraft, "but we soon learned to do the cockpit check from left to right using the mnemonics TMPFFGG for Trim, Mixture, Pitch, Fuel, Flaps, Gills, and Gyro." The Anson had an amazing reputation for safety. Fraser Gardner remembers seeing one "literally fly into the ground" in the low-flying area near No. 5 SFTS, Brantford, Ontario. Although the crash demolished the Anson, the instructor, student, and a passenger emerged from the remains and walked away unhurt. One night of extremely stormy weather saw an Anson instructor and his student end their trip "still sitting in their seats, side by side, high up in a large tree. The aircraft had completely disintegrated around them." Neither man bore a scratch, Gardner adds. But even Ansons occasionally surprised the unwary. Ted Johnston of Barrie, Ontario, took advanced flying training on Ansons at No. 9 SFTS soon after the school had moved from Summerside to Centralia, Ontario. He recalls his first flight there, a familiarization trip: "The instructor took all his students up at once. . . . We were shown run-up procedures, take-off, climbing and cruising speeds. . . . We were then shown how the Mark II Anson stalled with wheels down, flaps down, power off." After this came a demonstration of a full-power stall with wheels and flaps down. The instructor warned his students that this manoeuvre would be "a little more violent." "He wasn't kidding!" comments Johnston. "When the aircraft finally stalled, it flipped over on its back and went into an inverted dive. It took nearly three thousand feet to regain control. All of us were suitably impressed. We had a lot of respect for and not a little apprehension about our new aeroplane." Not until later did Johnston and his fellow students learn that their Anson, due to a fault in trim, had behaved with a violence uncharacteristic of the type.

Students loathed one feature of the early Ansons: the manually operated mechanism that retracted and lowered the landing gear, since they invariably had the job of turning the gear handle. It took more than a hundred and fifty revolutions of the reluctant handle to move the gear up or down. It may have been one reason why so many students forgot about their wheels. On the Anson a horn would sound to warn pilots if they descended too low with the gear still retracted; nevertheless, wheels-up landings were daily occurrences. Seldom did the students do much damage either to themselves or their aircraft. Many schools had airmen armed with Very pistols stand on the airfield, ready to fire warning shots if they saw aircraft approaching with wheels up. Bruce Betcher, an American in the RCAF, was at No. 5 SFTS, Brantford, Ontario, when he observed a fellow student, Australian Jackie Hallas, approaching in an Anson. It was a hot day, so Hallas had opened the bomb aimer's panel in the nose of the aircraft, helping to cool the cabin with a pleasant stream of air. As he levelled out prior to touching down, Hallas was more than a little startled when a "ball of fire" hurtled through the tiny aperture and whizzed past him en route to the rear of the aircraft. He quickly realized what had happened. He had neglected to put his wheels down, and the airman on the ground had, very properly, fired a warning shot. But instead of merely warning the Anson off, he had hit it, right in the open panel in the nose.

Hallas hastily wound down his landing gear. He landed, and, braking violently, skidded along the runway. "The rear door popped open and Jackie tumbled to the pavement, picked himself up, chased after the aircraft, re-entered it, and extinguished the fire," Betcher recalls. William Clarke of Greenwich, London, says that while he was an instructor at No. 33 SFTS, Carberry, Manitoba, one of his pupils landed an Anson without lowering his gear "and was surprised to find his windscreen being splattered with chips of wood as he sank lower and lower toward the ground. He was in fact grinding the wooden propellers down to about a quarter their normal size. Suddenly realizing what he had done, he opened up the engines and managed to lurch into the air and even turn round slowly and land in the next field in the opposite direction."

"Cherry" Cherrington of Birmingham, England, had been washed out of pilot training at EFTS and had remustered as an air bomber. While flying in an Anson, he was ordered to wind up the undercarriage – "a task that found me first using my left hand while seated, followed by my right hand while kneeling on the seat." Puffing after his exertions, he slumped back in his seat and picked up his maps, but by now he was completely lost – "and this," he comments, "on a map-reading exercise!"

The manually operated flaps of the Anson also caused problems. Former RAF student Ron Hunt remembers a staff pilot "who had a pronounced English accent" ordering his crew of two Australians to "Pump flaps!" Both thought he said "Jump, chaps!" whereupon they bailed out. Bill Cody, a Fleet Air Arm student pilot, found the "Canadian twang" of his instructor hard to follow. Approaching a low-flying area one day, the instructor barked,

"Check your gas." After asking the instructor three times to "say again," Cody cut the gas, believing that was what the instructor wanted. Fortunately, the instructor was quick to notice the gas cock being turned off.

While there was seldom much to choose between the elementary trainers of the period, students quickly recognized the pros and cons of the higher performance types. Jim Emmerson progressed from elementary training to No. 16 SFTS, Hagersville, Ontario, and, eventually, to the Mark II Anson, with hydraulic brakes, flaps, and landing gear. "But before we could step up to this luxury," he recalls, "we first had to endure the purgatory of the Anson Mark III." Essentially, this latter type was the result of equipping the old British-made Mark I Ansons with Jacobs L6MB engines as their original British Cheetahs wore out. The Mark III emerged before the considerably superior Mark II because of production delays. Emmerson recalls that the Mark III's brakes

> were run by a compressed-air bottle which chose the most inopportune times to exhaust itself. A cautious, creeping return to the hangar was required. The signal to ground crew to pump in more air involved putting the thumb into the mouth and blowing on it – a ludicrous performance more suited to baby carriages than a wartime training plane. At the pilot's left elbow was a huge iron lever of the type used to set a steamroller in motion. This was to release the undercarriage lock. The wheels then had to be cranked up – only 150 rapid turns if your arm held out. On landing, all this machinery had to be manhandled once again . . . while pumping flaps down by hand.
>
> Circuits and bumps – continuous landings and take-offs – were an exhausting nightmare of frantic pumpings and crankings. Our training began in August, and the

temperature in the confined cabin seldom fell below that required to bake bread. All this could have been endured with a patient, understanding instructor. But I drew one who was a Jekyll-Hyde personality – sweetness and charm on the ground, a raving fiend in the air. I was terrified that he was going to start beating me with a fire extinguisher for some flying lapse. If I managed to get something right, he would cackle like the village idiot and start clapping as if dealing with the dunce at kindergarten.

Emmerson began to wonder if it was all part of an elaborate psychological test, but on checking with friends he learned that their instructors seemed relatively normal. Eventually, much to Emmerson's relief, his instructor was posted away – possibly for psychiatric care, Emmerson thought. Jim Ruddell, a classmate of Emmerson's at Georgetown High School in 1939 and 1940, also experienced some problems with the Anson's braking system. When he set off on his first solo on the Anson, at No. 5 SFTS, Brantford, Ontario, he found himself in a Mark III, a type he had not previously flown. "I recall first having difficulty getting started," he writes, "then taxiing out with air brakes which I hadn't used, although others had told me about them." After take-off, Ruddell completed his circuit, then lined the aircraft up to land. But he found, as countless students did, that an Anson with only one occupant had a tendency to float. Sensibly, he abandoned that landing attempt and went around for another try. "This time the landing was OK," he says, "but, in attempting to brake to the taxiway, I didn't have the rudder coordinated with the thumb air brake control and ended up doing a 360. I told my instructor that I found the air brakes tricky and he was embarrassed to learn that I had never been checked out on that model."

While a fairly docile aircraft, the Anson needed a good deal of manipulation in flight. After flying more than a thousand hours

Brass: Robert Leckie (right) is generally credited with the idea of training Commonwealth aircrew in Canada, which grew to become the BCATP. He is seen here with Lloyd Breadner, chief of the air staff. Both men were fighter pilots in WWI. (DND PL21717)

Ken McDonald, one of the RAF instructors sent to Canada before the BCATP was created. He is seen here with an Oxford at Camp Borden, on a snowy day in December 1939. McDonald later commanded No. 7 AOS, before returning to Britain and flying a tour of operations with Bomber Command, winning the DFC. (Ken McDonald collection)

With the roller coaster of the Canadian National Exhibition providing a somewhat incongruous backdrop, RCAF recruits at No. 1 Manning Depot, Toronto, prepare to have their photos taken in uniform for the folks back home. (Ken Smith collection)

RCAF recruits soon found that marching was the service's favourite means of moving groups of them around, no matter what the weather. The smiles for the photographer are gallant but unconvincing. (Ken Smith collection)

En suite: At No. 1 Manning Depot, Toronto, the Agricultural Building of the Canadian National Exhibition accommodated RCAF recruits. The hastily installed facilities were hardly luxurious, and many recruits found the lack of privacy the worst aspect of service life.

Providing adequate supplies of potable water for BCATP schools - some of which consumed as many as 50,000 gallons a day - became a problem in parts of the prairie provinces, where water was highly mineralized. In a few cases, water reached the bases by various forms of transport, as here at No. 32 EFTS, Bowden, Alberta, in the summer of 1941. (Bob McBey collection)

First of many: Some members of the BCATP's first aircrew course at No. 1 ITS, the former Eglinton Hunt Club, Toronto, May 1940. John Simpson is seventh from the right. With few exceptions, the graduates of this course became instructors in the BCATP. (John Simpson collection)

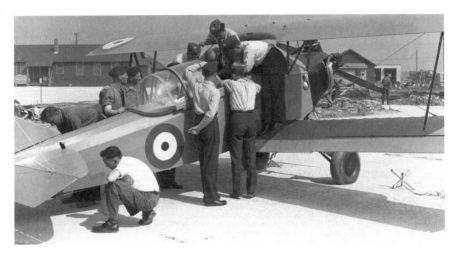

Eager student pilots examine a Finch elementary trainer before embarking on their first flights. The Finch, built by Fleet at Fort Erie, Ontario, and the Tiger Moth, a product of de Havilland of Toronto, were the principal elementary trainers in the BCATP's early days. (DND photo: PL1022)

Perfect flying weather for Tiger Moth No. 5804 from No. 31 EFTS, De Winton, Alberta. The unit was one of the RAF schools that later came under the control of the BCATP. (Tet Walston collection)

The monoplane Cornell trainer eventually replaced most of the biplane Finches and Tiger Moths in the BCATP. Designed by Fairchild in the United States and built by Fleet in Canada, the Cornell proved to be an excellent trainer, despite structural problems early in its career. (NAC PA 187754)

Ab initio: His smile reflecting the delight that most young student pilots felt on being selected for pilot training, Jimmy "Stocky" Edwards stands before a line-up of Tiger Moths at No. 16 EFTS, Edmonton, 1941. Edwards had a magnificent career as a fighter pilot, shooting down 15 enemy aircraft, earning the DFC and Bar and DFM. He rose to the rank of wing commander.
(J.F. Edwards collection)

Intrepid aviator suit: Trying out his flying garb at No. 2 ITS, Regina, is nineteen-year-old Jack McIntosh of Medicine Hat, Alberta. Graduating as a pilot, McIntosh went overseas and won the DFC, flying with 419 Squadron of Bomber Command.
(Jack McIntosh collection)

Briefing: Student and instructor confer before taking off for a training flight at No. 4 EFTS, Windsor Mills, Quebec. The aircraft is a Fleet Finch. (DND photo: PL2052)

Link: A student pilot concentrates on balancing the controls of a Link trainer. Many students found this predecessor of today's flight simulators far trickier to operate than real airplanes. (DND photo: PL5233)

CFI: Civilian flying instructor J. G. Cole of St. Paul, Min., gives student pilot Donald Henderson of Black Diamond, Alberta, a few last-minute instructions before sending him off solo. Henderson was killed in action in April 1943, flying a Wellington bomber. (DND photo: PL6104)

Dukes Mk. I and II: The identical and inseparable Warren twins from Nanton, Alberta, had remarkably similar abilities in the air. During training they sometimes substituted for one another. Both were nicknamed Duke. Later they flew together in combat, both winning the DFC. (Douglas Warren collection)

Harvard huddle: Basking in the California sun, factory-fresh Harvards await delivery to Canada for service in the BCATP. More than 20,000 Harvards of various marks were built for the Allied air forces. (Doug MacPhail collection from North American Inc., original via Jeff Ethell)

Admirable Anson: A pleasing study of Anson II No. 11433 from No. 15 SFTS, Claresholm, Alberta, summer 1943. Note the strips on the wing in the foreground, added to study the effect on the airflow of stalling. (John Simpson collection)

Designed as a light transport, the Cessna Crane served the BCATP well as a twin-engined trainer. Although some instructors considered it too flimsy for the daily punishment of training duties, it remained in service until the end of the war. (Phil Weedon collection)

Oxbox and Annie: At No. 1 SFTS, summer 1940, an Airspeed Oxford and Avro Anson are parked side by side on the flight line. Widely used in Britain, the Oxford equipped only a few schools in Canada; the Anson saw service throughout the BCATP. (John Simpson collection)

Battle: After a disastrous career in combat against the Luftwaffe, the Fairey Battle was relegated to training. It served the BCATP primarily on bombing and gunnery duties, although a few became advanced trainers because of shortages of more suitable aircraft. (Canadian Warplane Heritage Museum)

The Hudson, a military version of the Lockheed 14 airliner, served RAF Coastal Command for several years. Many journeyed to Britain in the hands of crews recently graduated from BCATP training schools. Gun turrets were installed upon arrival. (NAC PA 187755)

BCATP student pilot Cecil Durnin pulled off a remarkable landing in a field, minus one-third of his Harvard's starboard wing. He had been low flying over holiday cottages near Kingston, Ontario, and hit a cable. A Mr. Clark of Smiths Falls witnessed the incident and snapped this photograph.
(Cecil Durnin collection)

Harvard vs Yale: At No. 6 SFTS, Dunnville, Ontario, in June 1941, Harvard No. 2757 and Yale No. 3437 collided as they came in to land. Although injured, all three occupants of the two aircraft survived the crash. But Kenney Norfolk of London, Ontario, the Yale student pilot, died three years later in the crash of a Ventura in Iceland.
(Arthur A. Bishop collection)

Ground loop: The Harvard, with its narrow tailwheel undercarriage, could be a handful to land at times. Even experienced pilots sometimes ground looped the feisty trainer, often "wiping off" all or part of the landing gear, as seen here.
(Doug MacPhail collection from negative supplied by RCAF Memorial Museum, CFB Trenton)

Deforestation: During a photographic exercise near Souris, P.E.I., Anson I AX372 encountered a sturdy spruce, which nearly decapitated staff pilot Arthur Grime and caused serious damage to the aircraft. Sgt. Grime managed to get the crippled Anson back to base at Charlottetown and land safely. No one was hurt - but Sgt. Grime had some explaining to do.
(Tom Robinson collection)

on Ansons, instructor Jeff Mellon found the transition back to Tiger Moths surprisingly difficult. He says it took him about eight hours to re-acquire the necessary light touch to fly the elementary trainer again.

The Anson's fellow twin-engined trainer, the Cessna Crane, had been designed as a light transport and was forced into the role of trainer by the dictates of war. It performed adequately, but never seemed to earn the affection of students and instructors. RAF instructor Dennis Miller-Williams actively disliked the aircraft. So did Bob McBey, who became an instructor at No. 33 SFTS, Carberry, Manitoba; he felt the Crane was too flimsy for the grind of training. Chester Hull, the RMC cadet who had switched to the air force on the declaration of war, was also critical of the Crane. He became an instructor and a flight commander at No. 11 SFTS, Yorkton, Saskatchewan. A senior officer asked Hull for his opinion of the aircraft. Hull told him. An official letter of reprimand soon arrived on his desk. The air force didn't take kindly to criticism of its equipment.

Shortages of trainers in the early days led to the use of some odd types; the Fairey Battle, for example, which took so many of the RAF's most experienced crews to their deaths in May and June 1940. For a few months, the type became an advanced trainer at No. 31 SFTS, Kingston, Ontario. Paul Burden of Fredericton, New Brunswick, accumulated 114 hours on Battles at Kingston. He had progressed directly from the Fleet Finch, one of the most basic of military aircraft, to the Battle, which a few years earlier had been considered one of the most advanced. He recalls that the transition from the little biplane to the hefty monoplane was surprisingly painless, largely because of the Battle's excellent flying characteristics. But the examples in use at Kingston were, in the main, scarred veterans of the disastrous battles over the French and Belgian countryside. Many bore conspicuous patches over bullet and shell holes. All were "clapped-out," says Burden, and plagued by glycol leaks, with cockpit floors invariably slippery with the coolant, which not only leaked inside the Battles but also deluged windshields at inconvenient moments. Few spare parts were available for the Battles and much ingenuity was displayed by the ground staff in finding substitute parts or making them. The inmates of Kingston Penitentiary contributed to the war effort by manufacturing bomb racks and ring sights for the planes.

A few British Airspeed Oxfords also found their way to Canada for use on RAF service flying schools, but the type was never widely used in the BCATP. Similar in design to the Anson and Crane, the Oxford was in some ways a better trainer for the BCATP, since it was a trickier aircraft to fly and thus introduced students to the stern realities of handling operational types.

It was a different world at SFTS. You were back in the air force after your sojourn at a civilian-operated EFTS, with its good food, smiling female help, and generally relaxed atmosphere. The service feeling enveloped you the moment you entered the base. No casual sauntering with hands in pockets here; no cutting corners and taking short cuts across the grass between the buildings; no letting your hair grow back to something approaching pre-service length. Everything had to be done the air force way.

As at EFTS, your days at SFTS were divided between ground school and flying training. There the similarity ended. Previously, much training time had been spent learning the essentials of flying; now you had acquired a certain familiarity with the air and its mysteries. Your reactions in the air had become automatic, hands and feet co-ordinating smoothly without conscious involvement on your part. You no longer had to concentrate every fibre of your being on keeping the aircraft aloft. Now it was time for you to move on to the serious side of flying, with navigational and bombing exercises, night-flying, and formation flying. The whole process of training was bearing a closer and closer resemblance to *real* flying. The big new aircraft you now flew were the next best thing to combat types. Far speedier, far more powerful than those little puddle-jumpers at EFTS, they were enough to quicken any red-blooded pilot's pulse. Was it any wonder that you ached to get away on your own to dazzle the locals with a display of your skills and daring? All too aware of the dangers of combining youth, impatience, and high power, the COs of many schools maintained a display for the benefit of newly arrived students. Usually it was at the back of a hangar, and consisted of a melancholy tangle of twisted, blackened metal and wood: the result of low flying or unauthorized stunting, a reminder of what could and probably would happen if you tried to do too much too soon. "Thought he was a better flyer than he really was. Showing off to his girlfriend. We had to cut his remains out of the mess. In half a dozen parts. His girlfriend saw them all. Have a good look at the blood he splashed all over the place, untidy bastard. There. There. And there."

The practice may have helped to discourage unauthorized low flying and stunting, but it could never eliminate them any more than the sight of grisly traffic accidents has put a stop to speeding on highways. During 1941, 170 students and instructors died in accidents. The causes were easily categorized: forty deaths were the result of low flying and aerobatics, stalls caused thirty-seven, night-flying accidents caused thirty-one, collisions another twenty, fifteen were attributed to problems with instruments or propellers, with twenty-seven chalked up to miscellaneous causes. [4]

To its embarrassment, the air force had to admit that students were not always to blame. Group Captain J. G. Ireland informed a meeting at the War Staff College that in 1943 fifty-four of the fatal accidents at BCATP fields had been caused by low flying. But, he added, "The most serious aspect of this is the fact that, in the majority of cases, the aircraft was in the charge of an instructor or staff pilot." [5] It was a reminder, if one was needed, that most of the

BCATP's instructors were as youthful as their pupils and just as spirited. What's more, vast numbers of them were bored and frustrated with instructing; aerobatics and low flying were a good way to "let off steam."

At No. 5 SFTS, Brantford, Ontario, the student pilots often chose to fly over to nearby Hamilton on sunny, warm days. The reason, according to Fraser Gardner, who trained there in 1943, was that the nurses at one of the city's hospitals "used to sunbathe on the roof in the nude." Unfortunately, he adds, the identification numbers on the Anson's wings were all too prominent, and several incidents of low-flying were reported. A general order forbade pilots to fly over that area. But, he says, "the smart ones usually made one pass – from the sun – and kept going. By the time the girls got their eyes focused, the planes were long gone."

Jake Gaudaur, former football player and later commissioner of the Canadian Football League, was a flying instructor at No. 2 SFTS, Ottawa. On night-flying exercises, he was fond of zooming low over the southern part of the city, where his wife lived, to tell her that he would shortly be home. Approaching the apartment building, he would slip the propeller of the Harvard into fine pitch, producing an ear-splitting howl. When an editorial appeared in the local paper about the antics of the airmen from Uplands airfield, he decided it might be more prudent to phone.

Don Cooper, an FAA student pilot who trained at No. 31 SFTS, Kingston, in 1943, remembers that the punishment for low flying was to be sent to Trenton to do hard labour in the RCAF prison. A fellow student, a "street-smart Glaswegian who had spent eighteen months in the RN before transferring to the Air Branch," was enjoying zooming low over Rideau Lake in a Harvard when he cut through a power line. He managed to stay in the air, Cooper relates, but he "put the area out of power for three days." The hapless student claimed that his engine had cut out and then had miraculously picked up just before he hit the cable. No one believed him. A local inhabitant had witnessed the incident, and when interviewed, she said she hadn't heard the engine as the Harvard approached the cable, but heard it loudly after the cable was cut. Cooper comments:

> It was, of course, a peculiarity of the Harvard that you couldn't hear it until it had passed over you, then there was the god-awful howl. The Kingston CO made a deal with the pupil pilot. "If you continue to try your excuses, I will make sure you finish up in Trenton. Otherwise I will give you five days in the local limestone quarry, breaking rocks." He, of course, took the latter offer. He dug out a small cave as it was extremely hot, stayed in there, and we took him Cokes and cookies on our time off.

Another student who "did time" for breaking the rules was Cecil Durnin of Winnipeg. By August of 1942, he had more than 260 hours of flying time to his credit and was nearing the end of his stay at No. 2 SFTS, Ottawa. He had already completed several of his wings tests. In a few days he would be wearing the coveted badge on an officer's tunic, for his instructor had told him that he could expect to be commissioned. "At nineteen," Durnin remarks, "I was one happy guy."

But not for long. On August 23 he set off in Harvard No. 2592 to fly a cross-country exercise. Another student, engaged in the same exercise, flew in loose formation with him. The two Harvards arrived at Kingston. "Without giving it a thought, we 'beat up' the harbour," Durnin recalls, "making two runs at a number of ships. . . . We then climbed back up to altitude, perhaps two thousand feet, for the next leg. I remember thinking, 'You damn fool. Someone could report this and you'd be in real trouble.'" Durnin had already been reported for flying under a bridge on the Gatineau River. Fortunately for him, his instructor had intercepted the report and prevented it going to higher authority. "I was grateful to him," Durnin remarks, "and now here I was, at it again."

The two Harvards passed over a collection of holiday cottages beside a lake. A moment later they were at zero feet, their propellers leaving trails in the water. Roaring skyward, they turned and zoomed over the waving, swimsuited people below.

"To this day," says Durnin, "I don't know why, but I locked my harness in the 'back' position. . . . I was firmly fixed to my seat and could only move my arms, legs, and head. I'm sure this saved my life."

He glimpsed some wires. It was too late, he couldn't avoid them. With a violent *thump*, the Harvard hit the wires, wrenching to the right. Instinctively, Durnin applied hard left rudder and aileron, as well as full throttle and full pitch. Hold her down, he told himself. Build up your airspeed. Clear the trees. He made no attempt to turn. Straight ahead seemed the safest course until he could find out the extent of the damage. His instrument panel was little help, providing no information about airspeed, rate of climb, or altitude. His helmet had gone, jerked off his head and out of the open canopy at the instant of impact. Had his harness not been locked back, his body would undoubtedly have slammed into the instrument panel, probably stunning him. The Harvard would have hit the ground in an instant.

Still flying straight ahead, he cautiously climbed. But a shock awaited him. When he glanced to his right, he saw that about a third of the wing had vanished, taking with it the Pitot tube. No wonder the instruments weren't working. Perhaps six inches of the starboard aileron remained, providing a modicum of control. His left foot jammed the rudder pedal to its limit.

Briefly Durnin thought of bailing out and claiming that it had been engine trouble. But he remembered that evidence in the form of a large piece of his wing lay on the island – "and no doubt a number of cottages were without electricity or telephones." The immediate problem, however, was how to get the aircraft down without killing himself. No turns, he decided; he would put the Harvard down in the first field that looked smooth and big enough. When he saw a suitable expanse, he didn't delay but cut the throttle and "hit the ground going at quite a clip."

Durnin almost stood the Harvard on its nose before it came to a halt. He remarks that he smoked in those days and, with the Harvard at rest in the field, he lit the last cigarette in his package "and finished it in about five puffs." Soon curious locals began to drift onto the field. It was Sunday morning, and many had been on their way home from church when Durnin roared overhead, minus much of one wing.

But the air force wasn't impressed with Durnin's remarkable airmanship. Back at Uplands, he found himself on open arrest, essentially confined to barracks. Soon afterwards, the SPs (service police) placed him under close arrest and locked him in the guardhouse. In all probability, he knew, he would be washed out. He wasn't. The next day, the CO sentenced him to twenty-eight days in the "glass house" at Lachine, Quebec, the principal service prison in Eastern Canada. Durnin made the trip by train, handcuffed to a service policeman, the subject of many a sidelong glance from the other passengers.

At Lachine, Durnin occupied a six-foot-square cell, his days a dreary succession of marching, drill, physical training, hard labour with pick and shovel, and scrubbing his cell (the pail used for this work becoming a seat at meals). From time to time the inmates had to scrub the entire building. The guards would secrete a few cigarette butts and safety matches around the place, and then, after the cleanup, examine the refuse. If all the butts and matches were not accounted for, prisoners were searched in their cells. An illegal smoke was a major crime at Lachine. Durnin also remembers being hungry all the time. Lunch was the best meal of the day. "It was always a heaping bowl of stew," he remarks, "and it was good." The evening meal consisted of coffee and three slices of toast and jam. Like all prisoners, he had to remain silent unless spoken to.

During his incarceration, Durnin's class graduated. One of his former classmates, Douglas Cormack, now an officer wearing pilot's wings, came to visit him at Lachine. Cormack and Durnin had grown up together in Winnipeg. (Cormack died in October, 1943, when his B-26 crashed into the sea just off Nassau, Bahamas.)

Durnin was released from Lachine on Wednesday, September 16, having had his sentence reduced by five days for good behaviour. Back at Uplands, he graduated with Class 57 as a sergeant pilot. Two years elapsed before he was commissioned.

Ralph Green of Grandview, Manitoba, did his advanced training on Harvards at No. 9 SFTS, Summerside, P.E.I. He recalls that "seven or eight" students were engaged in solo night-flying one dark winter's night. When the daily train from Charlottetown put in an appearance, one of the students couldn't resist the temptation to fly ahead of it, then turn and head back, zooming low over the track. "About half a mile from the train," says Green, "he snapped on the Harvard's powerful landing light." The reaction was instantaneous. "The engineer had to know he was driving the only train on the island, but apparently, in a horrifying moment of panic, he envisioned an impending head-on collision and immediately applied full braking, scattering passengers and baggage from one end of the train to the other. Fortunately, no one was seriously injured."

Students found the powerful SFTS aircraft considerably more demanding than those encountered at elementary flying school. Some never managed to make the transition. When Len Morgan first took the controls of a Harvard, he discovered that things tended to happen rather more rapidly than he was accustomed to.

While the Harvard was a relatively easy ship to fly, it had a narrow gear and a high centre of gravity. When the tail dropped on landing, the wing blanketed out the tail group, rendering the rudder to a large degree ineffective. This unfortunate combination (also possessed by the Navy's Wildcat) made crosswind landings more than a little interesting. On the early Harvards which we flew, the rudder pedals were tied to the tailwheel – up to a point, that is. A sharp kick of either rudder would disengage the tailwheel, allowing the ship to swing around in its own length. This was a necessary feature for parking the ship in close quarters. On the runway it often worked against the pilot, however. You'd touch down in a cross wind and the nose would swing as the ship attempted to weathercock; you'd counter with rudder to straighten out. Jab it a hair too much and the tailwheel would come loose with a sickening snap. From this point onward things happened fast, unless you were quick with the brakes. The idea was to apply a few degrees of corrective rudder and hold it until heading was under control. No quick, deep kicks. Most of this we learned the hard way. [6]

Remarkably, a student named Harry Anderson at No. 2 SFTS, Ottawa, managed to reach the stage of solo flight in a Harvard without ever learning how to release the parking brakes. Alex McAlister watched as Anderson opened the throttle. Nothing happened. He thought the chocks were still in place. He waved them away. The ground crew saw no chocks, so gave him a thumbs-up.

Poor Harry was puzzled. By now, he realized that the parking brakes were on, but he didn't have the faintest idea what to do to release them. He kicked around a bit, then rammed open the throttle. The Harvard roared to life, then spun around in a tight circle. Obviously he had hit one of the brakes at random and released it, while the other remained as was. The fact that a mere tap on the brake pedal was all that was required was quite unknown to him. The bewildered Anderson hung on tightly. By this time, quite a reasonable sized group of interested onlookers had assembled to watch the show. Among the sea of faces, however, one emerged quite definitely and made way to the aircraft with considerable speed. This was accompanied by the person of the instructor in charge of Anderson. Harry was in the process of cutting the switches when the good instructor whispered several vicious sentences in his ear concerning the release of parking brakes . . . [7]

Another fellow-student at Ottawa, "Dunk" Flannigan, learned things about the Harvard the hard way while doing an instrument take-off. The instructor in the rear seat lined the Harvard up on the runway, then turned control over to Flannigan.

He opened the throttle and they roared down the runway. Everything seemed to be in perfect order. The aircraft left the runway and climbed straight as an arrow for a few hundred feet. Then Flying Officer Black noticed that they were climbing too steeply. "You're climbing too steeply, Flannigan," he shouted through the intercom; "push the stick forward!" There was a couple of seconds' silence from Flannigan, then he shouted back: "I can't move it, sir; it seems to be stuck." Flying Officer Black grabbed the controls and flipped the quick release for the hood over the back seat. This done, he looked back to see Dunk struggling with the stick which was still stowed away in the safety position for solo flying. Darned if the old Harvard hadn't done a perfect take-off with neither pilot touching the control column. [8]

Walter Miller trained at No. 8 SFTS, Moncton, New Brunswick, where both twin-engined Ansons and single-engined

Harvards were in use. The latter's tendency to ground loop resulted in a number of minor accidents. The Chief Flying Instructor (CFI) angrily informed all pilots that the next one to ground loop a Harvard would be washed out. "With a few minutes to spare one day, between classes," says Miller, "six of our class were watching the Harvards take off and land. Suddenly one Harvard ground looped on landing. The pilot was not hurt; he climbed out of his aircraft, threw his parachute over his shoulder and started to walk to the hangars. Our group said the chap was for the wash-out machine. Then the pilot came near and we all sprang to attention. The pilot was our CFI, Squadron Leader H. Bryant. He did not acknowledge our salutes and he looked quite mad." Miller adds that the squadron leader did not wash himself out.

The Harvard was a spirited aircraft and unforgiving of sloppy flying. John Macfie of Dunchurch, Ontario, who had successfully completed his elementary flying training without succumbing to airsickness, arrived at No. 2 SFTS, Ottawa, in February 1944 to begin advanced training. His class was greeted with the news that the air force now had a surplus of pilots. The washout rate was about to soar to unprecedented levels. In addition to that less-than-cheering news, Macfie found the training pipeline temporarily plugged up with pupils. To his chagrin, he was put to work for a month in the mess, a move he described as "criminal" in his diary. On April 3 he completed his first solo in the Harvard. But two days later he became a victim of the aircraft's testiness. He came in to land too fast and began to drift to the right. Hurriedly correcting, he touched down "in what I thought was the three-point attitude. . . . The right wing went down and swung to the right." Again correcting, Macfie now found the Harvard wrenching to the left. Before he could lift the port wing, it hit the ground. Fortunately for the youthful pilot, the aircraft "ballooned into the air." Hastily correcting, he overcontrolled. Slamming on full power, he managed to wrench the Harvard away from the ground, doing a steep turn and coming perilously close to turning it over onto its back. Fortunately he managed to bring the balky aircraft down with only minor wing damage. But the penalty was shattering: immediate dismissal from the course. Like so many washed-out students before and after him, Macfie was devastated. It seemed to him to be the end of the world. Now he would never wear pilot's wings on his tunic. It says much for the single-mindedness of these young men that a John Macfie would equate failure in flying training with failure in life itself. Not only had he let himself down, he felt, he had failed his entire family. Within hours he was writing his mother, telling her of the incident and admitting that mere survival was no consolation. [9]

Tet Walston, who had graduated from EFTS at De Winton, went to No. 32 SFTS, Moose Jaw, and became the pupil of a Battle of Britain veteran, Flight Lieutenant Draper, DFC, flying Harvards. Draper passed on some tricks of the fighter pilot's trade. "One was, when the enemy is breathing down your neck and is closing in the evasive turn, drop wheels and flaps suddenly. If he doesn't ram you, you'll lose him when you whip the wheels and flaps up again. The other was to increase speed in an evasive dive, put the prop into full coarse; it's like accelerating in a car."

The Harvard often proved more than a match even for experienced pilots. Len Morgan remembered an officer with combat time spinning into the ground near Trenton for no apparent reason. "The yellow Harvard hit flat and level, so that the fuselage was wrenched from its mounts and rested lengthwise along the wing. The pilot had jumped too late. I can see the reddened, grotesquely crumpled form as it lay where it hit, a few inches ahead of the muddy engine." [10]

In 1943 Jack Barnes of Toronto had already spent two years overseas as an RCAF instrument technician. He remustered as a pilot, flying Harvards at No. 1 SFTS, Camp Borden. Barnes admits that he "never felt too comfortable practising spins solo." He recalls that practice spins were always carried out above six thousand feet, with the stick and throttle being pulled back until the airspeed fell to 60 to 70 mph, after which it was a matter of inducing the stall by heaving back on the stick and applying rudder. The Harvard, Barnes says,

> would then flick over quickly and the descent would increase rapidly while the spinning intensified. After five or six spins, but no lower than three thousand feet, the stick was pushed fully forward and opposite rudder applied. This increased the speed of the spin which seemed to make it worse. The natural reaction is to pull back on the stick to stop the aircraft from diving at a steeper angle. But this is a mistake, since the aircraft would still be below flying speed and would stall again. It was the tightening of the spin and the higher speed with a steeper angle of descent that always worried me, although I knew it had to happen to provide the aircraft with enough forward speed when it came out of the spin. Remarkably, the Harvard snapped out of the spin very suddenly and it was a simple matter to pull back on the stick and apply reverse rudder to be flying straight and level again.

It wasn't only in the air that mishaps occurred. One unusually dark night at No. 6 SFTS, Dunnville, student Jim Morton of Burlington, Ontario, taxied a Harvard out in preparation for take off. Ahead he could see the beacon signifying the taxi position; a small red light on top of a wooden post. From this point he would turn and take off between the row of lights marking the runway. Everything appeared to be proceeding according to plan. Morton eased back on the throttle as the taxi position light neared . . . except that, too late, he was shocked to see that it wasn't the taxi position light. A tearing, a ripping, and a sickening crumpling sound punctuated this instant of realization. Horrified, Morton saw that he had trundled straight into another Harvard. Like an industrious buzz saw, his propeller had chewed its way through the tail unit and halfway up the tubular metal structure of the fuselage, reducing an expensive trainer to a tangle of scrap metal. Fortunately Morton was able to cut his throttle before his propeller reached the cockpit. The occupant scrambled out, shaken but unhurt.

An unfortunate set of circumstances had combined to precipitate the accident. What Morton had taken for the taxi position light was in fact the navigation light on the waiting Harvard's wing tip. The aircraft had completely blocked the static light from Morton's view, and the intensely dark night had compounded the problem, rendering the Harvard virtually invisible in the blackness.

Wilf Danby of Toronto also had an unnerving night-flying experience. On a frigid night in February 1942, he and a number of other students and their instructors flew from Camp Borden, Ontario, to the satellite field at Edenvale, a distance of about fifteen miles. "We found the airfield blanketed under more than a foot of snow," Danby recalls. "The runways and taxi strips had, however, been well plowed down to the pavement with the snow piled up in mounds along the runways to a height of three feet, higher in some places." The normal practice was for the night flying control officer (NFCO) to supervise the activities from a position at the end of the runway, communicating with the pilots by means of an Aldis lamp, and with a Very pistol in emergencies, since none of the Harvards was radio-equipped. All aircraft engaged in night flying also had to carry Very pistols and a supply of red and green cartridges. On this evening, Danby flew to Edenvale with his flight commander, who happened to be the NFCO that night. "After landing, he checked that there was an Aldis lamp in operation. There was, but there seemed to be some uncertainty about the Very pistol, so he removed the pistol from the aircraft – which was now minus an essential piece of night-flying equipment, and I was scheduled to fly it in a few minutes to start the night-flying program."

Danby took off and completed two circuits without incident, although he noticed that the pale blue lights marking the edge of the taxi strips were in many spots obscured by piled snow.

After the second landing, he rolled to the end of the runway, aware that other aircraft were waiting to take off. "I put on a burst of throttle and looked for the blue light that marked my turning point onto the taxiway," he recalls. But the heaps of snow obscured the light, and he was startled to find that he had crossed the taxiway at the end of the runway. A moment later he had become embedded in deep snow. "I realized that I represented a hazard to anyone attempting to take off. The proper procedure was for me to fire off a red flare from my Very pistol." But in this case it was impossible: the flight commander had removed the Very pistol from this particular aircraft! Danby pondered quickly. Jump out and get a message to the NFCO? Or remain in the cockpit and hope that someone was aware of the problem? He decided to abandon the aircraft. . . . "And in fact I had one leg over the side of the fuselage when I looked back and saw red and green navigation lights rushing toward me. Two thoughts flashed through my mind: no one knows you're stuck at the end of the runway, and . . . get down and pull your head in!"

During his take-off run, the pilot of the second Harvard glimpsed Danby's aircraft in the darkness. But it was too late to stop. Hunkered down in his cockpit, Danby "heard a fearful roar . . . followed by the agonized screech of tortured metal." Then silence. Danby – "with no little trepidation" – raised his head and looked over his right shoulder. He found that the fin and rudder of his Harvard had been sliced off; at the same time the propeller of the second Harvard had "chewed its way along a good portion of my starboard fuselage, finally coming to a stop no more than two feet from my head!" Neither pilot was hurt in the crash, although the two Harvards suffered severe damage.

For most students, solo flight at night was an awe-inspiring – and often terrifying – experience. John McQuiston of Toronto recalls that most of his course considered themselves "hot pilots" by the time they reached No. 16 SFTS, Hagersville, to learn to fly twin-engined Ansons. They soon found out how little they knew. "Even our instructors were pretty inexperienced," McQuiston admits. On his first solo flight at night, he was lined up with the runway in preparation for landing. Suddenly, to his alarm, he observed a car's headlights approaching. "Had there been no car," he says, "I would probably have landed on the main street of Aylmer. . . . Talk about inexperience!"

After his first solo cross-country at night, Lew Duddridge wrote:

> I felt as alone as I have ever felt in my life. Once over the airfield I looked down from two thousand feet at the green, red and amber lights that marked the runway. Never before had I gone much further than gliding distance from the safe confines of my home field. . . . This is where the nerves started to act up. It was a real challenge to keep to my flight path. There was such a temptation just to swing that aircraft around and fly in large, safe circles above those familiar lights on the ground. What did I want with heading off into the blackness? There was nothing on my instrument panel that was going to bring me safely home. . . . I had butterflies where I had never had butterflies before. But I fought it off, and the stars in the unfriendly gloom of night seemed to beckon me on. I couldn't resist." [11]

RAF student pilot Len Dunn was similarly awed on his first solo night flight from No. 35 SFTS, North Battleford, Saskatchewan, in the spring of 1942. "What a world! The black mass below, with its flickering lights, seemed to have no part in my life. What a sight! A full red moon dripped off my wing tip, constellations reached into infinity. . . . After that venture into eternity, life was never quite the same." Ex-RAF instructor Richard Taylor says, "Night-flying above the prairie in midwinter, usually in clear moonlight, is something I shall always remember. . . . It was a common thing, to break the long nighttime monotony, to fly against orders and beyond radio range to Winnipeg and try to read the name of the movie advertised at a cinema on Main Street."

Accidents continued to take their toll. Students, instructors, and ground staff suffered death or injury in many ways. Jim Northrup of Surrey, B.C., was an instructor at No. 3 SFTS, Calgary, in 1942, by coincidence commanding the same flight in which he had been a student a few months earlier. From his office, he saw Cessna Crane No. 7725 trundling along the taxi strip. Simultaneously, he caught sight of an airman walking across the strip, parachute over his shoulder, apparently deep in thought. To Northrup's horror, the airman walked straight into one of the Crane's revolving propellers. The blade caught him in the back, splitting him from neck to buttocks. Northrup says:

> A doctor and ambulance were there in minutes. But I knew it was curtains for the airman when I saw the doctor cringe as he pushed the boy together. I got the instructor and pupil out of the Crane and into my office. They were white as sheets. I got them some water as I thought they were going to pass out. I might add that I did not feel too well either, but being obliged to look after my instructor and student gave me something to do and probably made it appear that I was calmer than was the case. It was fortunate for the instructor that I had seen the accident, for the cry went up that he had been taxiing too fast, but I was able to state in no uncertain terms that this was not so.

The tragic incident occurred simply because the student was preoccupied and not looking where he was going.

Formation flying had its risks, particularly when inexperienced pilots were involved. One incident involved future

Canadian airline chief Max Ward, who spent some time as an instructor at No. 16 SFTS, Hagersville. He had taken up two students to introduce them to formation flying.

> I had arranged for two other aircraft to formate ... with our own Anson. Each aircraft had two students, and their instructions were to formate on our right- and left-hand sides and to hold their positions in that formation. While we were getting into position, a sudden movement to our right and behind us caught my eye. I looked again, and saw an Anson coming up on us, moving far too quickly. I told the trainee who was flying, as calmly and quietly as I could, "Okay, I'll take her," and grabbed the controls to hold the machine straight and level, just seconds before there was a tremendous crash and the Anson rocked and staggered in the air. The other aircraft had smashed into us, just behind the main cabin; its wing struck our fuselage and, as I glanced quickly out the window, I saw the stricken aircraft, with one wing torn away, flip over and plunge towards the earth. I was horrified, but there was nothing I could do, except try to get the aircraft back under control. The cabin was full of dust and debris from the crash and the aircraft was lunging all over the sky, but it resisted the temptation to go into a spin. The controls were sloppy under my hands – it was a bit like trying to fly through mud – but they did respond, and I was able to get the aircraft turned around and slowed down, and we limped back to Hagersville airport. Although it was not more than a score of miles away, it felt as if we had been flying forever; and when we finally got back to earth I discovered that the crash had torn fabric off the side of my plane, and it had wrapped around the elevators on the tail assembly, which explained why I found the aircraft so sluggish. The two students in the other aircraft were killed, and the rest of us survived

mainly by a fluke. The Anson was built of metal tubes, wood, and fabric, not a very sturdy craft by today's standards, but the original design had a gun turret just behind the cabin. In the training craft, the weight represented by this gun turret was replaced by a large block of concrete, and it was into this concrete that the out-of-control aircraft had crashed. A few feet forward, and none of us would have survived. [12]

Collisions were an ever-present danger; and with the number of aircraft in the circuit on a typical day at any BCATP airfield, the wonder is that there weren't more. Warrant Officer Danny Lambros of Wiarton, Ontario, was in the circuit at No. 1 SFTS, Camp Borden, one day in June 1942, riding in the rear seat of Yale No. 3358. One of his students was at the controls. Everything appeared to be in order for a routine landing – until a Harvard flew into the underside of the Yale. Such collisions often resulted in the two aircraft being instantly transformed into a tangle of burning wreckage. Lambros and his student were lucky. The Harvard's propeller slashed through the elevator and rudder cables which passed beneath the Yale's cockpits but left the basic structure undamaged. Lambros took over from his shaken student. The aircraft still flew, after a fashion, controlled by the ailerons and elevator trim, which continued to function; but for how long, no one could tell. Lambros quickly decided to abandon the aircraft. He ordered his pupil to bail out. No response from the front seat. He repeated the order. Still no reaction. Lambros hurriedly looked around for a suitable place to put the Yale down. Using his throttle judiciously, he was able to make a good approach to an open section of ground, part of the army training area near Borden. But as his speed slipped away, so did his all-important elevator trim. The Yale was virtually on its own for the last few moments of the approach. "If we'd been flying a Harvard," Lambros remarks, "we could have kept our wheels up, which would probably have made things easier." But the Yale had a fixed gear. It hit the ground with a teeth-rattling bang and bounced back into the air. Lambros could do little but wait to see what would happen next. He soon found out. One of the Yale's wing tips brushed the ground, whirling the aircraft into a cartwheel, a lunatic bounding over the ground, accompanied by a thunder of tearing, crumpling metal as an assortment of bits and pieces ripped away. At last, momentum spent, the Yale ended up on its back, a write-off. The occupants were remarkably lucky. The student pilot hadn't suffered a scratch; and Lambros's injuries were minor.

Most aircraft provided good visibility forward. The same could not be said of the view in other directions. George Stewart of Hamilton, Ontario, became acutely aware of this fact when he flew Anson II, No. 8229, solo from Centralia, Ontario, to nearby Hensall on a misty day in February 1943. "As briefed," Stewart recalls, "I began to do a routine exercise of practised forced landings at the designated farmer's field near Hensall. The visibility was quite restricted, and I assumed that mine was the only aircraft in the area."

He set up his forced landing approach, throttled right back, and began his final turn toward the field, descending normally. At about five hundred feet, he was startled to hear the sound of another aircraft. Then he saw it. The instant seared itself into his memory:

> Suddenly it was there, immense and vivid yellow, and I was crashing into it in one god-awful wrenching and tearing moment. My nose and starboard engine chewed through the other aircraft's port wing, and my starboard wing sliced into its tail. Time seemed to stand still – and then in a flash the other plane vanished and I was slamming into the ground, having landed totally by reflex and instinct. Now it was quiet, ominously quiet. I remember breaking out in a cold sweat. All I could hear was the whirr of spinning gyros. The smell of gasoline permeated everything. I could still picture the looming fuselage, huge and yellow, the number 7257 on its side.

Still dazed, Stewart scrambled out of the wreckage of the Anson, vaguely aware that his undercarriage had become wrapped up in the tail section. He ran to the other wreck a few yards away; it was hardly recognizable as an aircraft, having gone in upside down in a rolling motion.

> I remember sobbing with the knowledge that I had probably killed someone. Then I noticed an officer in the remains, bleeding profusely, swearing about his best blue uniform being ruined. I helped him free, after which I crawled inside, along the roof of the fuselage to the student trapped by the twisted controls, hanging upside down but, thank God, alive. Praying that the wreck wouldn't catch fire, I eased him into a more comfortable position and went back outside for help. By this time people were approaching the scene. I remember one kind lady bringing me a glass of milk and a sandwich.

Stewart still wonders how everyone survived the collision and why the aircraft failed to blow up in a horrific fireball. Unaware of each other's presence, the two Ansons had banked, virtually in formation, as they approached the same field. Neither saw the other; and the aircraft carried no radio equipment with which the pilots might have warned other aircraft of their position and intentions.

Two days after the incident, Stewart did an hour's dual with his instructor, then went off on his wings test. He graduated the following month with an "above average" rating. (Nearly fifty years later, Stewart and Jimmie Fairfield, the instructor aboard Anson No. 7257, made contact. Fairfield wrote: "I can't see how you could be in any way responsible. . . . One person alone in an Anson certainly lacked the starboard visibility enjoyed by having both seats occupied." Stewart flew a tour of operations on Mosquito night intruders with 23 Squadron, earning the DFC. He is still active in aviation.)

At No. 5 SFTS, Brantford, two Ansons collided as both approached to land, one aircraft simply settling on top of the other, no one aboard either aircraft having spotted the danger. Steve Puskas of Hamilton was a student at Brantford at the time and saw it happen. "The top Anson severed the tailplane of the lower one, slamming the control column into the face of the student pilot." Puskas remembers that the stricken aircraft performed a complete loop, rounding out just above the ground, while the second Anson went into the ground on its side. Both aircraft were write-offs, but, remarkably, there were no fatalities among the occupants, although the pilot of the lower Anson suffered the loss of several teeth, knocked out when the control column hit him in the face. Cecil Robson witnessed an almost identical midair collision at No. 17 SFTS, Souris, Manitoba, but in this instance, everyone aboard the two Ansons died. (Robson, a native of Truro, Nova Scotia, was taking advanced flying training after flying a tour of operations overseas as a wireless operator/air gunner with Bomber Command. He went through the BCATP twice as a student, later serving as a staff pilot.)

The commanding officers of BCATP schools studied accident statistics as diligently as bankers study market trends. By mid-1945, No. 17 SFTS had been accident-free for months. RAF instructor Dennis Miller-Williams was at the station at the time and remembers that the wing commander in charge of flying was very concerned. "His forebodings were based on statistics. We were told to be extra careful. as the number of hours for a major accident had been exceeded. He repeated the warning some weeks later." That day, two Ansons collided over the field, killing the occupants.

BCATP students got lost every day. So, occasionally, did staff pilots. Don Sinclair was at No. 31 ANS, Port Albert, Ontario, when one of his fellow staff pilots, Sergeant Jock Doyle, wandered across the U.S. border and landed at an airfield near Bay City, Michigan. When he pulled up to the apron, Doyle found his Anson surrounded by armed guards, "probably," notes Sinclair, "because the radio people couldn't decipher his broad Scots accent and took it for something more sinister." Eventually, Doyle convinced the Americans that Scotland was on the Allied side. Supplied with fuel for his Anson, Doyle flew back to Port Albert after an overnight stay.

On May 31, 1943, New Zealander Ron Anstey took off from No. 34 OTU, Pennfield Ridge, New Brunswick, for night circuits and bumps in a Lockheed Ventura. Also on board was Australian wireless operator/air gunner Gordon Carmichael. After a few circuits, Carmichael suggested that they fly further afield and listen to dance bands on the aircraft radio. "About an hour later," Anstey recalls, "when the time came to tune in for a bearing home to Pennfield Ridge, Carmichael had difficulty with the radio compass and couldn't get a bearing." The situation soon became serious. The airmen were lost and getting low on fuel. "The sky was black as ink and no lights anywhere. I was of the opinion that we were out over the sea. We flew around for some time until the petrol we had left would last only ten to fifteen minutes. I decided we would have to bail out soon and take our chances. Suddenly Carmichael saw a line of lights below . . ."

Luckily, the Ventura had been heard circling the small local airport at Augusta, Maine. Airport staff turned on the runway lights, and the Ventura touched down at 4:35 A.M., June 1, with two intensely relieved airmen aboard. A local aviation enthusiast, Ruel Hanks of 96 State Street, drove to the airport, then took the two airmen back to his apartment for breakfast. When Anstey returned to Pennfield Ridge, he was reprimanded for getting lost on what was supposed to be a local flight, his flight commander painstakingly drawing the cartoon character Pilot Officer Prune, the air force's pet blunderer, in his log book as a permanent reminder of his misdemeanour. Anstey later served with 98 Squadron RAF, flying Mitchells.

While they were not required to master the art of finding their way about the sky to the same degree as observers, pilots were expected to acquire basic skills. Walter Miller remembers that "navigation was difficult for me, but I worked long night hours at it. After our final ground school navigation written test, the navigation instructor came to me and said that before the exam he was sure I would fail it. He then said I had made a small error in addition which cost me five points, otherwise I would have made 100 per cent. Much sweating paid off for me."

Not every student was that diligent. Before he became an instructor, Jim Northrup went to No. 3 SFTS, Calgary, and frequently flew with a close friend, Bob Crosby. On navigational exercises, one flew and the other navigated – but, Northrup admits, they didn't apply themselves to the craft of finding their way about the sky as enthusiastically as they might have done. "We always seemed to have something more interesting to do than navigation," he says. Eventually, however, the time came for their night navigation exercise, which had to be completed successfully before graduation and the winning of pilots' wings. "The flight commander said it was our turn: a fairly simple cross-country, Calgary – Red Deer – Drumheller – Calgary, about a two-hour trip." Northrup clambered into the pilot's seat of the Anson; the flight commander occupied the right-hand seat, and Crosby's job was to navigate. Northrup says candidly, "We wandered all over the sky, and the only thing that got us back to Calgary was the oil field flares at Turner Valley." Unimpressed, the flight commander ordered the two to repeat the exercise the following evening, reversing the roles so that Northrup navigated and Crosby flew. "Unfortunately I was no better a navigator than Crosby," says Northrup. The next night saw an equally undistinguished performance, saved only by the invaluable presence of the Turner Valley facilities and its illuminations. "Fortunately we were off for two or three nights," Northrup recalls, "so Crosby and I used the time to do a little brushing up. Soon I was back in the pilot's seat, the instructor in the other seat, and Crosby in the back. Now we were really paying attention. We hit Red Deer bang on and turned for Drumheller. Everything was proceeding perfectly." Then, disaster! "Over Rosetown, about ten minutes from Drumheller, both engines quit, out of gas! I pushed the flare switch, but nothing came out. The instructor gave the order to bail out." The trio landed safely, but the Anson was destroyed in the subsequent crash. The farmer in whose field the airmen landed provided them with a lift into Calgary. "By this time," remarks Northrup, "the flight commander was so fed up with us, he said we would have to learn to navigate somewhere else!" In fact, Northrop and Crosby got their wings without ever passing the night navigation test.

Frank Phripp, who had found the Harvard more than a little daunting when he first encountered it at SFTS, recalls an "Instrument Cross Country" exercise; unfortunately he and his assigned instructor had a "somewhat strained relationship," Phripp remarks.

> My instructor . . . was in the back seat while I flew under the hood from beginning to nearly the end. We flew a few courses west and north from Dunnville to the Owen Sound area and then back south. I was happy enough with the task throughout the two to three hours of the exercise, but as we came south over the rather indistinguishable farmland, I noticed more questions and some concern in my instructor's voice . . . he didn't know where we were; *but I did* . . . The hood snapped open, unannounced, to hit me on the head, as was the pattern with this inconsiderate SOB, and he directed that I get a pinpoint and set course for Dunnville. I did this silently and deliberately within a couple of minutes of receiving the instruction. We came over Dunnville exactly on ETA [estimated time of arrival]! After landing, my instructor

quizzed me about my setting course so quickly. I
shrugged an answer, but didn't tell him that when he hit
me on the head with the hood, the first thing I saw was
the Nith River bridge in New Hamburg that I had
worked on in the CNR. Piece of cake, that exercise!

Weather was a perennial problem. Ted Johnston remembers
that at No. 9 SFTS, Centralia, there was "some friction between
the navigation section, which was responsible for our cross-country navigation flights, and the flight commander, who had to get
us through the required number of hours." In the winter, Centralia's location on the shore of Lake Huron, in the "snow belt,"
added to everyone's difficulties. Although frequent, says Johnston, snow squalls were, for the most part, "spotty and short-lived.
Flying could be stopped because of poor visibility, then in a short
time the sun would come out for the rest of the day. The cross-
country flights were especially hard to complete. While the navigation section tended to say, 'If the weather gets bad, come back,'
the flight commander usually advised, 'It's likely the snow will just
be local, so don't give up too easily. You're not going to have perfect weather when you get overseas.'" Johnston ran into a squall
one winter's day and made his decision early: he was going
through.

The snow squalls made it a challenging navigation exercise to get to Toronto. It was clear, there. Feeling quite
proud of myself, I started back to Centralia. After about
ten minutes, the snow really started. I didn't want to fly
completely on instruments, so I picked a good railway
track going in my direction and flew lower and lower to
keep it in sight. I figured I'd soon be through, and then I
could get back on course. Intent on keeping the track in
view, I was suddenly jolted when I saw trees *higher than
me* out the side window! I had pulled the classic mistake
and had forgotten about the rising topography near

Belfountain, my location. It was still snowing, so, with my heart pounding, I went on instruments and climbed out of there. At what I considered a safe height, I took stock, having used up one of the nine lives students are supposed to have.

The rule at Centralia was, when lost, to fly west until you reached the shore of Lake Huron. If it ran north and south, you were to turn south until the shoreline curved west, then turn inland, look for the emergency field at Grand Bend, and head home from there.

After what seemed a lifetime, it started to lighten up, and then the snow stopped and the sun came out. There was the lake shore running north and south; not only that, but there were Tiger Moths all around me. I was at Goderich. I flew south to Grand Bend, then back to base.

Everyone else had turned back, and when I reported to the navigation section, I got hell. Was I trying to kill myself? Wreck the aeroplane? What did I learn about navigation on a flight like that?

But back in the flight room, the flight commander made it all worthwhile. "I'm sure glad I have one trainee who isn't a fair weather pilot," he said. I was nineteen, and who has any sense of fear at that age? Just luck!

Although the Battle of Britain had ended in victory for the RAF, it had been a desperately near thing. Just how near neither the Germans nor the British public knew – and the Ministry of Information preferred it that way. What most people didn't understand was that, although the Ministry of Aircraft Production under Lord Beaverbrook had achieved minor miracles of aircraft production and had regularly replaced losses within hours, the same couldn't be said of the men who flew them. The RAF was fast running out of pilots – and hasty training sessions for Bomber Command and Fleet Air Arm pilots to teach them the elements of fighter tactics was strictly a stopgap measure. "Of Dowding's problems most serious by far was his shortage of fighter pilots. The Battle of France and Dunkirk had claimed 435 killed, missing and imprisoned. At mid-June [1940] Dowding was 360 pilots short of his full complement of 1,450. Canadian Prime Minister Mackenzie King's refusal in 1937 to cooperate in the Empire Training Scheme now hit the RAF at its most critical hour," wrote Peter Townsend in his history of the Battle of Britain. [13] The simple truth was that the RAF had gone into battle with too few pilots – and, incredibly, it appears that the flying schools in England were "working at only two-thirds capacity" during the battle, "following peace-time routines, being quite unaware of the grave shortage of pilots in Fighter Command." [14] Had the negotiations between Ottawa and London not ground to a virtual standstill for month after month, more trained pilots would have been ready to meet the *Luftwaffe* when it came. Had the Battle of Britain been lost, had the Germans successfully invaded, the self-satisfied King might well have found himself cast, to his undoubted indignation, as one of the villains of World War II, remembered by the world as being among those responsible for Britain's downfall.

Throughout the war, the training of aircrew was of necessity a compromise between the ideal and the practical. In peacetime, most pilots had about three hundred hours' flying by the time they reached their squadrons, where they spent several more months under the tutelage of their flight and squadron commanders. The pressures of war didn't permit such a leisurely pace. Airmen had to be turned out rapidly, but not so rapidly that they would be of little use on their operational squadrons. The numbers-over-quality approach had been tried in World War I, with disastrous results. But now, with the enemy on Britain's doorstep, the risk had to be taken again: training courses had to be shortened to get pilots into combat with all possible speed. Since all parts of the Empire Air Training Plan had to produce a virtually identical "product" – that is, an airman ready to go to work for the RAF – changes in training curricula in Britain had to be reflected in similar changes in Canada and elsewhere. Service flying training was cut progressively, from sixteen to ten weeks.

Essentially, the cuts were made by shifting responsibility for such aspects of training as gunnery practice and formation flying, to operational training units (OTUs, formerly known as group pools) in Britain. The change was far from popular among the operational commands, for they considered themselves overworked already; they favoured even more training at the flying schools, not less. The OTU concept had important benefits, however. At OTU a pilot or indeed a whole crew could start getting to know and fly operational types, although the majority of available aircraft were battle-weary veterans a long way past their prime. Many were downright unsafe. Herb Hallatt of Windsor, Ontario, was at Catfoss OTU in the summer of 1941 when Coastal Command crews were learning to operate the recently introduced Beaufighter, which had a daunting reputation as a "hot" airplane. He recalls that one by one the pilots and observers in his hut were killed on night-flying training, until he was the only survivor. Scores of newly trained crews died in crashes involving obsolete, war-weary Bomber Command aircraft which should have been scrapped but had to be kept flying because the OTUs had nothing else to fly.

In Canada, EFTS courses were reduced from eight weeks to seven; SFTS courses went from sixteen to fourteen weeks, then to ten. The distinction between the intermediate and advanced phases of service training had now all but disappeared. The SFTSs handled fifty-six students every twenty-four days, instead of the forty every four weeks that had previously been the norm.

It all added up to a gratifying increase in the numbers of pilots being produced by the BCATP. Originally estimated at 3,196 for the 1940-41 period, pilot output actually amounted to 7,756. The numbers were pleasing and made good reading in official reports, but there is little doubt that the quality of training had suffered. Students, particularly those going overseas, had to cope with long periods of inactivity between SFTSs and OTUs, and their newly acquired skills deteriorated proportionately.

While the emphasis was on flying during the latter part of the SFTS course, ground school still occupied much of the students' time – and failure in the classroom could mean the humiliation of a washout even at that late stage in a man's training. Sometimes, however, high marks were just as important to the trainers as to the trainees. Don Cooper, a FAA student pilot at No. 31 SFTS, Kingston, recalls that some three weeks before the end of his course, there was much concern about forthcoming tests in ship and aircraft recognition. In the case of the former, the RN petty officers solved the problem. "They were supposed to have taught us ship rec," says Cooper, "but they were lousy at the job, and they knew it – but they also wanted to keep their cushy jobs at Kingston; far better than corvettes in the North Atlantic! They were supposed to walk around and supervise during the exam. They did, with their hands behind their backs in true naval fashion – and flashed cards identifying the ships as they passed." In the case of aircraft recognition, Cooper himself was the class hero. He had spent some time in the Air Cadets before enlisting and had never experienced any difficulty in distinguishing Messerschmitts from Mustangs, Heinkels from Hurricanes. He and his classmates devised a plan. In the first part of the aircraft recognition test, pictures of various aircraft were flashed quickly on the screen, and the students had to mark them as friend or foe. "I sat in the front row. As the aircraft were flashed on the screen, I would lean forward to write if it was friendly and stay upright to write if it was foe. The second part of the exam," Cooper relates, "was to name the aircraft. We practised lip-reading for two weeks before the exam. I lip-read to the next guy, and he passed it back by the same method. The course average was 85 per cent – the highest ever, apparently – and no suspicions, only congratulations."

Throughout the war, errors in aircraft recognition resulted in untold numbers of casualties. The authorities worked diligently on the problem of distinguishing friendly and enemy aircraft. Sketches and detailed plans appeared in countless publications; scale models dangled from ceilings in every air force establishment. Training journals carried stories of poor aircraft recognition and its dire consequences. Typical was the tale of the Mustang pilot who spotted a German Junkers 88 bomber flying at three hundred feet near the French coast. The Mustang pilot fired one burst, then realized the other aircraft was a Beaufighter not a Junkers. Fortunately his shots had missed. But another Mustang appeared. It too mistook the Beaufighter for an enemy. It opened fire. And missed. The Beaufighter got away and landed safely back at base, where its pilot reported that he had been attacked by two Messerschmitts but had beaten them off.

For all student pilots, the high point of the entire training process was the wings parade, at which they would be presented with their cherished pilot's badges. All the hard work had paid off; they were about to become members of the élite – air force pilots!

But before the great moment arrived, every student had to undergo his "wings test," a demonstration of his prowess before a testing officer, often one sent from Trenton for the purpose. It was the last of the major hurdles for every budding pilot and a source of much tension among all but the most phlegmatic. You *could* do it all; you had proved the fact dozens of times – but the trick was to do it with some super-critical Trenton type in the right-hand seat (or in the back cockpit in the case of a Harvard). At No. 33 SFTS, Carberry, Manitoba, one of the testing officers, Hubber by name, was known as Hubber the Scrubber. Leonard Cheshire's former air gunner, Richard Rivaz, now a student pilot, had Hubber for his wings test. Having been told that Hubber paid particular attention to pupils' composure, Rivaz resolved to be as casual as if he were taking a taxi ride. Hubber ordered him to do a series of climbing turns, then steep turns and maximum rate turns. Approaching for a simulated forced landing, Rivaz nearly overshot the field and had to side-slip violently before managing to pull off a good landing. Taking off again, Rivaz headed for base, whereupon Hubber cut the port engine, necessitating a right-hand circuit of the field, since it is always dangerous to turn in the direction of a dead engine. Successfully maintaining his air of imperturbability, Rivaz carefully set up for his single-engined approach. Everything seemed to be in good order, and he began to congratulate himself on the successful completion of the test. "We were about twenty feet from the ground, and fifty yards or so from the runway, when a red Very light shot across our nose. 'Your *wheels!*' Hubber shouted. I had forgotten them . . . completely forgotten all about them!"[15]

In spite of this gaffe, Rivaz got his wings.

Jim Northrup was selected to be the first in his course to undergo the wings test. He remembers "a few butterflies in the belly" at the prospect:

> I met the testing officer, Flight Lieutenant Rowlings. He was a tall, well-built chap, quite pleasant, and every inch a professional pilot as you would expect from the testing flight. It was a beautiful clear, sunny day with just a little wind from the west bang on the east-west runway. After I did the usual outside inspection, we sat in the cockpit [of a Crane] for a time while he asked me various questions. Then he told me to start up and taxi out to the runway. After an hour and five minutes, we were back where we started. He completed his report in the aircraft. Just before we got out, he said, "That was fine, Northrup." I knew I had passed. It was now just about lunchtime, and

as my course mates and I walked to the mess hall, I was deluged with questions about what the examining officer had put me through. I had passed my instrument wings test four days before, so now they all knew I was going to get my wings. I answered every question as best I could and assured them that the examining officer was a decent chap and that they would make it.... After lunch the next chap selected for testing was Ben Flatt, the most popular student on the course, just out of high school. . . . We were all outside the flight room as Ben, Rowlings, and Flying Officer Weeks, a Link trainer instructor, who was going along for the ride, walked out to the same aircraft I had flown in the morning, Cessna Crane No. 8701. We watched the take-off. About an hour later we spotted the aircraft flying east to west at about one thousand feet, right over the runway. At the same time a Fleet Fort, a single-engine aircraft from the wireless school that operated at No. 3 SFTS, was flying west to east on exactly the same flight path. They saw each other at the same time and tilted their wings to miss. However, the Fort cut about three feet off the starboard wing of the Cessna. The wing started to disintegrate. The Crane made two complete somersaults and hit the ground. All three in the aircraft were killed.

Northrup adds that the scene is as clear to him today as it was half a century ago. "I believe had they not seen each other at the last moment, they might have made it past each other without colliding. The Fort landed safely."

For several days prior to "Wings Day," the students strutted about the parade square (or in a hangar if the weather was inclement) in meticulous ranks, rehearsing the entire ceremony from beginning to end. When at last the great day arrived, they donned freshly pressed and drycleaned uniforms, with dazzlingly white aircrew flashes in their wedge caps. Their faces acquired self-satisfied smiles that refused absolutely to budge. This was their moment.

Sometimes it was the CO, sometimes the mayor of the local town, sometimes a duke or duchess, sometimes a visiting hero of the calibre of Guy Gibson, sometimes it was Billy Bishop himself who did the honours, pinning the wings on chests that seemed at times to be in acute danger of bursting with pride. Bishop revelled in such moments. His son, Arthur, recalled: "He believed in bands, parades and lots of publicity. The fact that he himself was the focus of most of the publicity did not faze him. It was in a good cause." Bishop officiated at "as many wings parades as he could, and deliberately made them as colourful as possible. Bands blared and flags fluttered as columns of blue-uniformed airmen marched."[16] Bishop always had a word or two for the graduate pilots, congratulating them on their accomplishments, wishing them luck in their future careers, often telling them that they had to kill the enemy but it wasn't necessary to hate him, for he was just doing his job as they were doing theirs.

At some RAF schools, the wings parades were considerably more matter of fact. Although a member of the RCAF, Bert Houle of Massey, Ontario, went to No. 32 SFTS, Moose Jaw, Saskatchewan, an RAF school, which had a few vacancies at the time. When the time came for graduation, Houle says, an instructor awarded the pilots' wings by standing on a chair and tossing them to the individuals as their names were called.[17]

The wings had been won. The flags had flown. The bands had played. Now the students had new worries. How many of the class – and, more important, *who* – would be deemed worthy of receiving the king's commission? Usually the top third of graduating classes became officers, the rest remaining as sergeants. The system was bitterly resented, both by the uncommissioned students and, in many cases, by their instructors, for large numbers of them still wore sergeants' stripes while their former pupils donned the snappy blue of a pilot officer. "This created an immediate class distinction between two groups," declares former instructor Chuck Appleton. "Those with commissions now had a smart new uniform, much superior perks in their messes, and their pay was $6.25 per day as against $3.95. And yet we were all expected to carry on and do exactly the same job. . . . It was particularly hard to see my students graduating as pilot officers and I was still an NCO ... We should all have graduated at the same rank and earned promotion from there on." Appleton was later commissioned, but he points out that "generally speaking, those of us who graduated as sergeants were usually about a year behind the others as far as promotions were concerned."

The problem of commissioning persisted throughout the history of the BCATP, despite the generally more liberal policies of the RCAF in comparison with the RAF. It affected all trades. In February 1945, when Norman Whitley's class at No. 1 CNS, Rivers, Manitoba, was about to graduate, the war in Europe had clearly entered its final days. Everyone expected to be discharged right after graduation. Or almost everyone. One member of the class was commissioned and it cannot have been without significance that he was the son of a senator. Had he earned exceptionally high marks during training, there might have been reason to believe that he won his commission on merit and not through political connections. But his marks were no better than average. Knowing the right people was a signal asset in the BCATP as in most other endeavours. But sometimes influence led only to tragedy. Instructor Peter Wrath saw a crash at No. 1 SFTS, Camp Borden, in which a student nose-dived into the ground while attempting to land at night. The unfortunate student had previously been washed out by a testing pilot, but his father, a prominent politician of the day, had insisted on his son being given another chance to earn his wings. The politician succeeded only in killing his son.

Although the top third of the class often became officers, other factors sometimes influenced the final decision. John Clinton of Burlington, Ontario, became an instructor at No. 16 SFTS, Hagersville. He remembers one of his students, an African, who turned in an exceptional performance during training. His flying, ground work, and deportment were first-class. But when Clinton recommended him for a commission, a senior officer tried to dissuade him, strictly on the basis of the man's colour. Clinton refused to alter his recommendation. The African got his commission.

CHAPTER FIVE

The Other Aircrew Trades

Avro Anson I

"Why argue who is most important? Without the pilot you're grounded. Without the WAG you're deaf... without the air bomber you're harmless... without the navigator you're blind . . . without the air gunner you're defenceless. Each job is vitally important.... It's the *team* that counts!"
— wartime poster

The Blériot monoplane that took to the air one August morning in 1914 closely resembled the famous craft that had wobbled bravely but uncertainly across the Channel five years before with M. Blériot himself at the controls. Now the world had gone to war, and aeroplanes were needed for sterner purposes.

The pilot of the Blériot, a young officer named Captain (later Air Chief Marshal) Philip B. Joubert de la Ferté of 3 Squadron RFC, turned and waved a greeting to a BE2A biplane piloted by

Lieutenant G. W. Mapplebeck. The two airmen had orders to patrol the left flank of the British Expeditionary Force (BEF) as far as Nivelles. But the cloud thickened, the wind grew stronger, and it became increasingly difficult for the airmen to remain in visual contact. With a final flick of gauntleted hands, they went their separate ways.

Thus ended the RFC's first reconnaissance operation in World War I. Over the next two weeks, dozens more were flown. It was Joubert himself who, on August 23, spotted grey columns of troops on the move. *Germans!* Joubert turned for home as scores of rifles crackled below. A bullet drilled a hole in the brass fuel tank in front of the cockpit. He thrust a gloved finger in the hole and, to his relief, the engine clattered on. He continued his journey, landing back at his home field and delivering the news of the enemy's advance. He may have saved the BEF, for Sir John French, its commander, was unaware of the enemy's attempt to encircle his army. Not without reason did Sir John later commend the RFC for the excellent information it provided.

Since the first hesitant days of flight, aircraft manufacturers had done their best to sell the military on their products. In 1905 the Wright brothers had suggested – unsuccessfully – to the U.S. Army that their invention might be useful for reconnaissance and artillery spotting. American generals, like those in Europe, considered aircraft nasty, noisy things that frightened horses – which was reason enough to have nothing to do with them. Not until war broke out did they realize how valuable aerial reconnaissance could be. In the British army, the success of the observation flights of August 1914 made such an impression on the general staff that, as far as they were concerned, observation was *the* reason for the RFC's existence; even the fighter aircraft that later came to symbolize the air war over the western front became known as "scouts," and, in the army's opinion, their *raison d'être* was to protect the observation aircraft from the attentions of enemy machines.

During that war and afterwards, manufacturers produced scores of aircraft of essentially the same type: single-engined two-seaters (the egregious Wapiti was among the last of the line) with one position occupied by a pilot who flew the machine, the second occupied by the really important fellow, the observer, a multi-talented individual who navigated, dropped bombs, took photographs, spotted artillery barrages, sent and received messages, and, when necessary, defended the aircraft with his machine gun(s). In the German Army Air Service of World War I, there could be no doubt about the relative importance of the pilot and observer: the observer was the captain, the pilot a mere chauffeur. It is one of the minor ironies of aviation history that in the RAF's World War II bombers, practically all of the observer's time was spent navigating to and from the targets at night and he had less and less time to *observe*; in fact, no member of the crew, with the possible exception of the wireless operator, observed less. Most observers in Bomber Command spent their wars enclosed in their curtained-off cubicles, intent on their charts and sums, communicating with their crewmates by means of the intercom, and looking outside only when dropping bombs or on the rare occasions when they took star shots. Their successors, the navigators, would be even more confined.

On the outbreak of war in 1939, virtually no RCAF airmen wore the "O" half-wing badge of the observer. Most of those in the RAF were veterans of the Great War and had been shunted into administrative jobs; many had become squadron adjutants. Large military aircraft of the inter-war period seldom carried observers. The tiresome chore of navigation was usually handled by the second pilot. In daylight he relied on map-reading. On rare night flights, it wasn't too hard to find one's way from one brightly lit city to another. Such a casual approach to navigation may have been adequate for peacetime; it wasn't for war. The RAF entered World War II confident that its bombers could attack targets in daylight – *precision* attacks, a factory here, a power station there.

But the crippling losses of the early months soon persuaded the Air Staff to switch the raids to the hours of darkness.

It proved infinitely easier to issue the order than to carry it out. Bomber Command's crews were, in the main, incapable of finding their targets in the blacked-out cities of wartime Europe. They hadn't been trained for the job, and they carried no special navigational equipment to help them. Their shortcomings soon precipitated a crisis at the highest levels of government. Churchill had made the bomber the cornerstone of Britain's offensive in the early days of the war. It was the only weapon Britain possessed that could hit the enemy at will. It *had* to succeed. The air marshals had assured everyone that, given enough bombers, they could bring the Germans to their knees, shattering their morale and destroying the nation's ability to wage war. Indeed, once the bombers had done their work, an invasion of the continent might hardly be necessary, they declared. The British had pumped vast amounts of money into the production of heavy bombers and the training of aircrew to fly them – principally through the BCATP. But squadrons of bombers that can't find their targets might just as well stay on the ground.

After being officially ignored for decades, it must have been a source of wry amusement to veteran observer/navigators to find the Air Ministry declaring in 1942:

> In many respects the air observer has the most responsible and exacting task in a bomber aircraft. . . . Mentally he must always be on the alert. . . . He must estimate and plot the course, be able to take snap readings, judge weather conditions, look out for ice and keep alternative objectives and landing grounds in the back of his mind. . . . He must show a marked ability to handle figures, and be sufficiently skilled in signals to take a portion of work off the wireless operator. Above all, he must never make mistakes. In most types of aircraft he has

considerable exercise in getting astro sights and taking up a bomb aiming position, and this may involve temporarily disconnecting his oxygen apparatus at a great altitude. Thus his fatigue is great. He is a wise and considerate pilot who appreciates the difficulties of his air observer.[1]

When the dignitaries applied their signatures to the BCATP agreement in December 1939, the need for observer/navigators had already been recognized. Ten air observer schools (AOSs) were planned. Between them, they would graduate 340 observers a month, more than four thousand a year, a gigantic number compared to what had gone before. The majority of AOSs were built on municipal airfields, sharing the facilities with elementary flying schools. As in the case of the EFTSs, civilian aviation companies ran the schools, supplying the staff pilots and mechanics, the administrators and the service personnel. The RCAF provided facilities, equipment, and instruction.

No. 1 AOS at Malton, Ontario, began operations on May 27, 1940. It had been organized by Charles R. Troup and W. Woollett of Dominion Skyways, a bush flying company operating primarily in northern Quebec. Troup and Woollett formed a subsidiary, Dominion Skyways Training Limited, and took on the operation of the AOS on a non-profit basis. (This latter provision was later the source of some acrimony. In 1941 two MPs, Howard Green and future prime minister John Diefenbaker, accused the company of profiteering. Feelings ran high at the parliamentary investigation, at which there were many "unpleasant and unjust innuendoes," according to Troup and Woollett, with Diefenbaker, in particular, acting "in a manner reminiscent of a prosecutor movie lawyer in a murder case.")[2]

Troup, known to everyone as Peter, was a former RAF pilot who had emigrated to Canada in 1928. He became the first manager of No. 1 AOS, and he found it a challenging job. Due to the cancellation of shipments of Battles and Ansons from Britain after the fall of France, the school had only fourteen of its complement of twenty-four aircraft. In an attempt to relieve the shortage, the Department of National Defence purchased several used aircraft from U.S. airlines, including two Boeing 247s, twin-engined airliners similar in appearance to the DC3 and unsuitable in every way for the work at hand. Troup wrote that "after careful consideration and thought we have rendered our reports on these aircraft and put them in the back of the hangars in the hope that a garbage remover will take them away one day." [3]

Troup and Woollett transformed the commercial company into a training company, hiring additional personnel at salaries and wages similar to those at Dominion Skyways before the war. The private enterprise aspect of AOS operations had a totally unexpected benefit for RAF observer student Jock Lovell. He recalls that at No. 5 AOS, Winnipeg, all the students received gift packages from the civilian contractor who looked after the school's facilities. The canteen had shown a healthy profit over the past few months, and this was his way of thanking the students. "A completely unnecessary and quite endearing gesture," comments Lovell. On the other hand, Allan Turton of Ottawa found to his dismay that "food at Ancienne Lorette [No. 8 AOS] was terrible and got steadily worse. We complained to no avail. We refused to eat the greasy half-raw eggs and bacon at breakfast and made ourselves stacks of toast. The civilian staff's answer was to hide the toaster."

The pioneering work of Troup and Woollett paid off in the development of the other observer schools. Typical was No. 7 AOS, Portage la Prairie, Manitoba, some fifty miles west of Winnipeg. It commenced operations in April 1941 and shared the airfield with No. 14 EFTS. Portage was a standard BCATP school, with eighteen acres of buildings and a triangle of landing strips, originally gravel, later concrete, each about 3,000 feet long. Two single hangars, 112 by 160 feet, and two double hangars, measuring

224 by 160 feet, comprised the flight line. Portage Air Observer School Limited, a subsidiary of Yukon Southern Limited, ran the school. (Yukon Southern and six other similar firms were purchased by Canadian Pacific Airlines in January 1942. Between them, they operated observer schools at Winnipeg, Prince Albert, and Portage la Prairie, as well as Ancienne Lorette and St. Jean, Quebec. Leavens Brothers operated the school at London, Ontario, and R. H. Bibby, a well-known bush pilot, ran the school at Chatham, New Brunswick.) The civilian chiefs were well known in the world of bush flying: Cy Becker, general manager, Matt Berry, operations manager, and Scotty Moir, chief pilot.

At Portage la Prairie, as at all other AOSs, a small contingent of air force personnel looked after the RCAF's interests. The officer placed in charge of No. 7 was Ken McDonald of the RAF, who had come to Canada from Britain just before the outbreak of war. He had spent the intervening months first at Camp Borden, Ontario, teaching navigation to pilots, then at Trenton, where he served under the command of Frank Miller (later chief of the defence staff), followed by postings to Rivers, Manitoba, and No. 4 Training Command Headquarters, Regina. By now, McDonald wore the two rings and "scraper" of a squadron leader. And he had a bride; in September 1939 he had gone on a blind date organized by Jean Lay, a niece of Mackenzie King, and had met Ruth Craig of Toronto. The couple married the following year.

McDonald flew a twin-engined Dragonfly biplane from Regina to Portage. His first glimpse of the airfield was somewhat daunting. It was early spring, and the field was frozen solid. Soon afterward the thaw set in, and the place became a vast expanse of gumbo. When McDonald flew as a passenger with Scotty Moir in an Anson, he took a professional interest in the way the ex-bush pilot avoided the gumbo: "He put on full flap, held it with power just above the stall, side-slipped off the last hundred feet, and landed on the hangar apron." McDonald had the highest regard for the civilians at Portage: "They [Becker, Berry, and Moir] and a surprisingly small staff flew and maintained the Ansons and looked after the infrastructure without any fuss but with great efficiency." What McDonald calls the "Canadian blend of service discipline and training with the expertise and management capabilities of the entrepreneurs who ran the civilian-operated schools" was in his opinion the key to the success of the BCATP. He recalls an incident at No. 7 AOS which typified this attitude: "One of the civilian pilots let a (non-flying) service instructor take over the controls of an Anson, and the latter made an ass of himself, flying it low over Portage. Technically it was a court-martial offence, but Matt Berry dropped by the office, smiling in his slow, quizzical way, and said, 'Ken, why don't you have your guy in and make him wish he'd never been born, and I'll do the same with my guy. No one was hurt, they won't do it again, and there'll be no paper.' That was the spirit I remember from those days in Canada: getting on with things without a lot of fuss." He recalls that relations between pupils and instructors were far less formal in Canada than in England, "where the staff had a separate ante-room in the mess and fraternizing with the students was discouraged."

At ITS, recruits told each other that anyone who displayed too much ability in mathematics was sure to be tagged as a potential navigator. Had you made your living as a bank clerk, accountant, bookkeeper, or comptroller before entering the air force? Your fate was sealed if you had, they said. Vernon Williams of Hamilton, Ontario, had worked as an accountant at The Steel Company of Canada when he volunteered in December 1940. Thus it was perhaps not surprising that at ITS he soon failed the Link trainer test – "co-ordination not good enough," he was told – and found himself tagged for observer training. He didn't mind; he would have taken any job the air force offered, as long as it was aircrew. Charles Onley of Toronto was a student accountant when he enlisted, and although he did well on the Link, he was picked for observer training. Joe Foley of Ajax, Ontario, had completed three years of his pharmacy internship when he joined the army in September 1939. He found himself at Petawawa, near Ottawa. Dysentery broke out in the camp, which had no medical officer. With his knowledge of medicine, Foley was hastily set up in a tent and given the temporary rank of medical sergeant. Later he was sent to Peterborough, Ontario, to supervise the construction of a camp hospital. His rank had now risen to warrant officer. When two medical officers arrived, Foley stayed on to run the place. "However," he says, "I didn't join the forces to be a hospital administrator." He favoured the air force, which at that point in the war seemed to be the service making the biggest contribution to beating the enemy. In November 1941 he volunteered for the RCAF and, to his surprise, was transferred without further ado. He had no idea it was so simple to move from one service to another. From the dizzy heights of warrant officer rank, he plummeted to the depths as he donned the uniform of an AC2. Good at mathematics, he was soon tagged as observer material.

The trade of "observer" embraced a variety of disciplines in those early days of World War II; little had changed since 1918. Although his primary function was to navigate, "the observer would leave his maps when near the target and take up a position over the bombsight, where he would remain until the bombs were gone. This work done, he could then resume his plotting duties, and pick up his position from a rough estimate of distance covered between the time of his leaving the target and recommencing the plot."[4] But in the event of a fighter attack, the observer became a gunner: "Fighter attacks were not only possible but very probable, and against a comparatively slow moving bomber they might come from any or all angles at once. There must be no blind or unprotected parts of the bomber. This meant that every man must, in that eventuality, be prepared not only to see what was coming, but also to ward off the attack. The observer was therefore to be trained to man whatever guns the bomber carried, whether fixed, free or in turrets."[5]

At AOS, the student navigator faced a classroom curriculum every bit as demanding as that faced by pilot trainees. It all started with the student learning the mysteries of navigation – the art of "accurate approximation" – with lectures on maps and charts, the basic tools of the navigator, accompanied by a welter of esoterica: grid references, abbreviations, scale, latitude and longitude, true north, magnetic north, magnetic variations, great circles and rhumb lines, Mercator's projection, meridional parts, conic projection, the international polyconic and its characteristics, and navigation lights. Then came a barrage of data on the application of the triangle of velocities – wind direction and velocity, course, and airspeed – to find track and ground speeds, followed by such arcane subjects as plotting dead reckoning (DR) positions during and after frequent alterations of course, finding wind velocity by obtaining drifts, calculating critical point and time on a flight in a straight line, calculating wind velocities by the reciprocal course method, intercepting an aircraft given its track and ground speed, time, apparent time, mean time. Then there were aircraft compasses, magnetism, hard and soft iron, compass swinging and adjusting . . .

The budding navigator also studied meteorology and weather maps, pressure distribution, depressions, anticyclones, the wind in relation to isobars, the variation of wind with height, water vapour in the atmosphere, stability and instability of the atmosphere, precipitation, types of cloud formation, effects of topography on fog and wind, thunderstorms and squalls, as well as the various instruments of his calling: the airspeed indicator, altimeter, airspeed calibrator, gyroscopics, sighting instruments, the loop aerial, direction-finding (D/F) facilities and systems, plus the ins and outs of air photography and a daily diet of aircraft recognition instruction. It was usually a week or two before the budding navigator got near a real airplane.

Student observers discovered a startling difference between the theory and practice of navigation. In a comfortable, well-lit classroom it all made perfect sense, in essence just a rather elaborate geometric exercise. In order to get from point A to point B you studied your chart, plotted your required track, and, aided by the given wind direction and velocity, determined your true course. Making your adjustments for magnetic deviation, you could confidently set off on your theoretical journey. In practice it was vastly more complicated. For one thing, the compasses and airspeed indicators aboard the aircraft of the period were only relatively accurate – and few pilots flew really precise courses. In classroom exercises, winds obligingly blew from one direction at a steady velocity. In the real world they had a nasty habit of shifting their direction and strength minutes after take-off.

On one of his first air navigation exercises, Joe Foley found out how utterly confusing the whole business could be in the uncooperative sky. On landing, he found that his log of the trip contained only four lines: "Four lines of the turning point information which I had drawn on the ground, nothing more. The log should have been full of pinpoints, sun shots, and D/F positions." Despondent, convinced that he had failed and would never earn his observer's wing, Foley sat on the grass and attempted to analyze what had gone wrong. Before take-off he had plotted the flight, using the mathematical compass, dividers, and a ruler, as he had been trained. But when he took to the air, the whole exercise became totally bewildering. Nothing seemed to work as he had expected. He found himself unable to calculate the wind variations and the changes of ETA at the various turning points. Why? If he could do it on the ground, why couldn't he do it in the air? Not until later did he realize that it was a case of his anxiety to do well turning into tension, and tension turning into a kind of numbing panic. Everything he had learned was forgotten. "In less than ten minutes after take-off, I had no idea of our ground position, D/F signals seemed to provide unplottable fixes, and the sextant bubble danced around so erratically that a reading was impossible. Though I realized that the pilot undoubtedly knew the position of every blade of grass below, it gave me little comfort. The exercise was a navigational disaster." The majority of students shared such experiences.

At AOS, most students spent eighty to over a hundred hours in the air. Courses started with simple local pinpointing and wind-finding exercises, learning to identify features of the ground and to estimate distances. A typical exercise for student navigators was as follows: "The pilot of the aircraft will fly on a predetermined course from the base aerodrome for ten minutes, at the end of which time the pupils will pinpoint their position. The pilot then flies back to the aerodrome whilst the pupils work out the wind velocity by track and ground speed methods. This procedure is carried out four times to complete the exercise before the pilot lands." On such air exercises, two student observers were usually carried: "Set course for objective sixty to eighty miles away, mete-orological wind being used. The 1st Navigator will check wind by determination of track and ground speed made good. 2nd Navi-gator will practice checking of drift by Course Setting Bomb Sight and Tail Drift Sight. Navigators will reverse duties at turning point." Triangular cross-country flights involved one student "making reconnaissance report on details seen" while the other concentrated on navigation. At first such exercises were based on legs of sixty to eighty miles; later the distances increased to about three hundred miles.[6]

Soon the air exercises became more complex: "The pilot is to fly on any course from the base aerodrome and is to make several large alterations of course at irregular intervals of not less than ten minutes. He is to inform the 1st Navigator of these changes, giving him the Magnetic Course steered, and the time flown on each. After one hour's flying, the 1st Navigator is to set course for the base aerodrome." Long cross-country flights were more involved: "A flight of 250-300 miles is to be carried out in the morning to another aerodrome. The aircraft will be refuelled. The Navigators will change duties and carry out a further flight of 300 miles before returning to the base aerodrome. Checks of position and ground speed will be made by use of D/F W/T [direction finding wireless transmission] bearings."[7]

Interception exercises demanded calculations of moving objects other than your own aircraft. For example, intercepting a train "running to known schedule, returning to base after interception or ten minutes after estimated time of interception. Exercise to be repeated, 1st and 2nd Navigators reversing duties, the second interception exercise being made of a different objective."[8] Photographic exercises involved such challenges as taking shots of vertical stereo-pairs of five pinpoints, making one line overlap cross-country between given pinpoints, taking five oblique shots of different sized objects, and creating a photographic mosaic of an area. The way such exercises were graded puzzled some students. Revie Walker of Blairmore, Alberta, trained at No. 3 AOS, Regina, and embarked on a navigational exercise one very warm spring day. During the flight, the second navigator was taken ill and couldn't complete his work assignment. Walker saved the day by giving him his log to copy, which he did. The marks came as a surprise. "He got an 8," Walker says, "and I got a 7!"

While a small error on one leg of a journey might result in the aircraft being only a mile or two off course, the accumulation of errors could rapidly transform a minor problem into a major disaster. Every instruction to the pilot had to be checked and rechecked. John Saqui of Glasgow came to Canada to be trained in the BCATP at No. 33 ANS, Mount Hope, Ontario. He recalls that, "As part of navigation training, we were required to do a square search. The work . . . is quite hectic, as the navigator is required to keep supplying the pilot with new courses. Consequently the u/t [under training] navigator has his head down, with little or no chance to look out. We were instructed to fly from Mount Hope to just north of Toronto, make a turn, and commence to search for a small pond. After several alterations of course, following the textbook procedure, I was politely asked by the pilot to have a look outside and try to locate the pond." Saqui glanced down to find the vast expanse of Lake Ontario beneath him. Some pond! Going back over his log, he realized that, "I had instructed the pilot to turn to starboard when north of Toronto instead of to port." Bob Caton, an RAF student navigator from Swanley, Kent, made no mistakes when instructed to map-read his way to a certain lake in Manitoba. Still he couldn't find it. He searched for some time before he realized that the bone-dry salt bed below *was* the missing lake, or what was left of it after a period of drought.

Frank Covert, the former lawyer who had received an over-abundance of gravy at manning depot, found the training unrelenting at No. 1 AOS, Malton, Ontario. "We studied awfully hard on the ground, and were soon flying in Avro Ansons, first as second navigators, then as first navigators, then on night flights. I don't believe I have ever felt so incompetent. Seldom did I suffer to such a great extent from the fear of failure. It spurred me to work harder and harder, but no flight ever satisfied me and some were nightmares! Once I was 'flustered' and nearly panicked. Once I was lost! My logged-flight marks went from bad to fair to good to bad! I almost despaired."[9]

Dead reckoning – the process of deducing your position from your last known position, using mathematical calculations involving airspeed, course, wind direction, and wind velocity – was the cornerstone of navigation taught at AOSs. Navigators learned to consider such concepts as D/F radio and astro (celestial) observations as mere aids to check on the accuracy of their D/R calculations. The era of total reliance on electronic navigation was still a few years off. The AOSs used "synthetic training," or simulated air exercises. In the Synthetic Dead Reckoning Trainer (SDRT), students sat in cubicles complete with most of the basic instruments used for navigation in the air. Classes were split into teams consisting of pilots and navigators. They communicated by intercom just as in the air. Briefings preceded exercises, with complete information on weather conditions, route, speed, altitude, and so on. To simulate night flights, the cubicles could be darkened, and the navigators worked by desk lamps, much as they would aloft.

The Celestial Navigation Trainer (CNT) – usually known as the "silo," because of its shape – was a more elaborate simulator. The CNT enabled students to practise navigating by the stars, and it could be set up to display any part of the globe or sky. Complete crews took their positions with the simulated heavens above and the appropriate section of terra firma below, and a smoke blower added cloud effects. Crew members communicated with one another by intercom. The CNT operator could introduce new winds and an assortment of other weather conditions during the "flight," while instruments in the control room below recorded the track of the aircraft.

Simulators had their place, but none could duplicate actual flight. Many a student navigator, although he may have done well in the classroom, found it appallingly difficult to work in the cramped, odoriferous, and uncomfortable conditions aloft, bombarded by incessant noise, usually freezing or frying, and constantly bounced around in the lurching and swaying that was a part of almost every flight. It was like trying to work on calculus in an evil-smelling truck with broken springs, careering along a pot-holed road at top speed. Ron Cassels was a student at No. 7 AOS: "The instructors demanded neat and accurate charts which required a lot of concentration as the Ansons bounced all over the sky. At that time farmers had a lot of summer fallow fields and each created a strong updraft which would lift the plane fifty to a hundred feet. You would be jammed down in the seat and your nose would almost hit the chart. The green fields and water had a downdraft and the plane would drop, causing you to lift off the seat."[10] Australian Don Charlwood recalled that his first navigation exercise "left us in a state of confusion and utter despair. Gravity plucked gut and skin, engine noise assaulted us, vibration and turbulence shook our bodies incessantly, clear thought seemed scarcely possible." [11] In those days, astronavigation played a significant role in the education of every student navigator, although many later wondered why, as only a relatively few navigators made much use of the science on ops. Cassels remembers spending "hours and hours" learning astronavigation, taking "hundreds of sextant shots," but he says he never used a sextant overseas. RAF trainee Gerry Powell received even more instruction in astronavigation than most of his fellow students. He was one of a few budding observers sent from Canada to learn navigation with Pan American in Miami, Florida. The airline, with its network of overseas routes, relied almost exclusively on astro, a fact reflected in the instruction syllabus. It was time largely wasted, Powell says, for he never had occasion to use celestial navigation in his subsequent air force career. (He was in Miami before Pearl Harbor, so the RAF trainees were disguised as civilians and had to endure the muggy heat of southern Florida wearing heavy British suits.)

Some Ansons had astrodomes on top of their fuselages from which the students could "shoot" the stars. But many Ansons weren't so equipped. Allan Turton trained at No. 8 AOS, Ancienne Lorette, near Quebec City, where a major problem in the winter was keeping warm while getting your work done. "The navigator's heating system was a half-inch tube that blew a draft of hot air across the table. We always plugged it with chewing gum, because it blew your maps and notes away. The open roof hatch in place of an astrodome was murder at forty below. To operate a sextant you had to remove the heavy mitt and glove from your right hand, leaving only a silk glove for protection. Your face took quite a beating out in the slipstream. We soon learned to 'cook' our sextant shots in cold weather." He remembers a New Zealand student navigator who had a finger amputated after "holding his sextant out of the Anson's roof hatch at forty below zero."

Some of the Ansons had no heating at all, and students rated winter astronavigation exercises in such aircraft as cruel and inhuman punishment. Astro took time, lots of it. You couldn't simply open the hatch, take your shot, and quickly close the hatch again. You had to endure the agonizing cold while you did your sums. Even then the results were often disappointing. You might be as much as a hundred miles off course by the time you had completed your calculations. The ceaseless battle against the clock took its toll on some students. If only the damned airplane would stop for a moment to let a fellow collect his thoughts, warm up his fingers, and check his figures . . .

To add to the student's discomfiture, training aircraft tended to be repositories of a selection of unsavoury odours: fuel and oil, hot metal and dope, plus sour reminders that not all the previous occupants had been able to keep their stomachs under control. The combination of such fragrances and the violent bouncing of the aircraft often led to feverish sweating, a conviction that one's limbs had turned to lead, and, usually within minutes, a violent upchucking. Allan Turton recalls that at No. 8 AOS, student observers often took cardboard boxes with them on flights. If they vomited into the boxes during the flight, they "invariably flew low when returning to base and 'bombed' farmhouses with the full cartons."

Most students found that daily flying gradually conditioned them and they overcame their tendency to airsickness. Some experienced greater difficulty in solving the problem. Ron Cassels remembers vomiting "sometimes for three hours straight. . . . It got so bad that I would start feeling sick on the ground before the plane started to move. . . . One day I was flying with Harvey Carmichael and we were both sick and the pilot reported us, with the result that we were told that we would be washed out. We pleaded that we be allowed to continue and got a stay of execution. The officer in charge was very definite that one more report and we would cease to be aircrew. Harvey and I made a two dollar bet, the winner would be the first one who made a trip without getting sick. A couple of trips later we flew together and both of us got through without being sick. It was a big hurdle." [12]

Others couldn't overcome the problem. Alan Turton remembers a Belgian student who "was very anxious to get into action" but whose tendency to airsickness became progressively worse. At first he began to get sick after a few minutes in the air, then he started to feel sick while taxiing; finally, the mere act of walking out to the aircraft was enough to induce the heaves. He washed out. Many navigators went through their entire air force careers surreptitiously carrying suitable receptacles on every flight – surreptitiously, for airsickness usually resulted in being grounded, particularly in the latter stages of the war. Charles Rawcliffe, an RAF student, was about halfway through his course when he was grounded for that reason. "I landed from a night exercise and the officer in charge of our flight asked me if I had been sick." Thinking he would be given something to cure the problem, Rawcliffe admitted that he had indeed been sick. "Unbeknown to me, all MOs had received a directive to weed out all personnel suffering from airsickness, as too many were getting their wings and sergeant's stripes but were unfit for ops." Instead of receiving medical treatment, Rawcliffe found himself reduced in rank to AC2 and on his way back to Britain, where he remustered as a wireless mechanic. Revie Walker recalls one warm spring day in 1941 when, after an interception exercise, the civilian staff pilot decided to indulge in some beating-up of farm fence-posts. Not usually susceptible to airsickness, Walker had consumed a lunch of roast pork and ice cream shortly before take-off. It reacted violently to the aerial gyrations, and he threw up. On landing, the pilot ordered Walker to report to the MO. All too conscious of the danger of letting officialdom know of any tendency to airsickness, Walker told him he would do as ordered, but not until he had reported the pilot's low-flying; whereupon the pilot decided that perhaps it might be best to forget the whole thing – a good thing for the air force, for Walker subsequently flew three operational tours with Bomber Command.

New Zealander Ron Mayhill was informed in no uncertain terms that airsickness was nothing but a question of mind over matter, something that could and should be suppressed by will power. Drink as little as possible before flying, chew gum to keep the mouth fresh, and occupy yourself during the flight; such were the instructions of Mayhill's MO at ITW (Initial Training Wing, the equivalent of ITS), Rotorua, New Zealand. "Anybody who was repeatedly airsick could not do his job properly, and not only would he be letting himself down, he would be endangering the rest of the crew." The MO's advice appears to have been sound advice, for Mayhill, who had suffered badly from travel sickness as a child, had no trouble in the air during his training or on his subsequent tour of operations with Bomber Command, during which he earned a DFC.

Instructors constantly demanded accuracy and speed, and not without reason. Dave McIntosh recalls the difference between his training days and ops in Europe flying Mosquito intruders:

An awful lot of my navigation trips during training were from Malton to Sharbot Lake. Between pinpoints, usually water, there was time to relax and work out drift, star shots, and the like; in other words, go by the manual. But in Europe it was madness. We did quite a few low-level "day rangers," and you had to follow your track second by second or you were lost. In Canada there might be a railway track every half hour; in Europe there was at least one every thirty seconds. I used two pieces of nav equipment: the *Gee* box to get us across the enemy coast at (presumably) the right place, and my pencil marked off in inches – an inch being one minute or four miles (luckily, we cruised at 240 mph or it would never have worked out). The pencil provided direction, time, and distance on my knee-borne map.

The staff pilots – a patient, long-suffering breed – spent their days and nights flying wherever the students told them to fly, making use of their detailed knowledge of the locality to find the way home if necessary. "The poor student might get lost," says ex-RAF navigator Hugh Redfern, who trained at the Central Navigation School (CNS), Rivers, Manitoba, "but not the pilot! There were times when they could have been so helpful to the struggling u/t nav, but as long as the aircraft was not in danger, they wisely said nothing.... They let us learn our lessons the hard way!" Allan Turton remembers how the American staff pilots at No. 8 AOS steered clear of the U.S. border because they were "afraid that if they flew over New York State and had to make an emergency landing, they would be drafted into the American armed forces. Consequently, if they found a student navigator putting the plane on a course that was getting too close to the border, they would alter course to the north without telling the student. The poor trainee would then wonder how his plotted wind velocity could change in ten minutes from 135/40 mph to 015/70 mph." Alan Helmsley recalls that as a student observer, he was flying with an American staff pilot close to the U.S. border. Suddenly the pilot realized what date it was: July 4! A moment later, Helmsley says, "the Anson was down on the deck and we were skimming along the border. Farmers leapt off their stacks, horses ran off in all directions, and my plot 'went for a Burton.' I could only sit back and enjoy the ride." Nova Scotian Ken Fulton remembers the prairies as a good area for navigational training. With the flat terrain and "the section line grain elevators (with names on), it was almost impossible to get lost." He admits, however, that "some of the pilots did not have much faith in the student navigators and sometimes altered course and height to check on the name on the nearest elevator. The best pilots (as far as the student navigators were concerned) were the bush pilots . . . [who] had the ability to know where they were at all times." But even staff pilots had to learn. Revie Walker remembers landing one day at Regina in an Anson. The pilot turned off the active runway, whereupon the aircraft promptly trundled to an embarrassing halt as both engines died, out of fuel.

"Our pilot was not too experienced, but tried his best," Walker comments generously.

A staff pilot's job was generally rather dull – although the air force did its utmost to boost the morale of the pilots themselves by asserting that their work required a "high degree of skill." Notwithstanding the kind words, the majority of staff pilots were far from content with their lot. A few turned to alcohol. Others indulged in more spectacular ways of enlivening the long and tedious days. At Ancienne Lorette, the staff pilots did a lot of low flying, says Allan Turton: "Favourite stunts were putting the Anson in a vertical bank and flying between the twin steeples of the Ancienne Lorette village church, flying under the Quebec City Bridge, making a low pass beside a ship on the St. Lawrence below the level of the main deck, and flying over Montreal at night low enough to look up at the illuminated cross on Mount Royal." Tom Robinson, a wireless operator from Northern Ireland who had started his RAF career as a boy trainee at Cranwell, encountered a high-spirited staff pilot at No. 31 General Reconnaissance School (GRS), Charlottetown, P.E.I., "a stocky, jovial sergeant named Arthur Grime, who lived to fly and looked forward to the day when he might be transferred to Mosquitoes or Beaufighters." One morning Robinson was assigned to fly with Grime in Anson Mark I AX372. Orders were to head out to several D/R positions over the Gulf of St. Lawrence, make a landfall at East Point, P.E.I, then go on to Souris, where, Robinson recalls, "there would be a turn into the bay and a gentle climb, so that our student navigators might photograph the harbour before returning to base."

It was a routine trip, one that Grime, the pilot, particularly enjoyed, because it gave him an opportunity for low flying and, Robinson adds, a chance to "scare hell out of the natives as well as his crew . . . when we arrived at East Point, Arthur usually went bananas. Skimming over the waves, he would swing into Souris harbour at high speed, apply maximum power, and climb steeply as our unfortunate navigators struggled with their clumsy cameras. Directly above the town, he would cut power, roll the sturdy crate on its side, and dive for a tree-covered point some distance away."

On this occasion, Grime headed for the ground with notable panache as bits of rubbish floated weightless in the trembling fuselage. Robinson glanced out at the instant of a "sickening shudder" and all he could see was trees. The pilot poured on maximum power and "there was a thunderous bang. . . . In a matter of seconds, the floor was completely covered with spruce boughs, and a branch six feet long and one inch thick rested on a ledge after having passed Arthur's left ear on the way in."

Staggering, *sans* windshield, with a spruce limb thrusting out in front like a leafy lance, the Anson struggled to acquire a little altitude. A frigid gale screamed through the narrow confines of the cabin. Robinson had to fight to make his way up front, where he "viewed the carnage with amazement. The main problem at this point was a lack of forward visibility brought about by spruce foliage lodged in the aircraft's nose." Robinson and the students attempted to reach outside and tear off some of the branches, but the effort accomplished little. Ninety per cent of the spruce remained firmly embedded in the aircraft.

The twenty-minute flight back to Charlottetown became tense indeed as the airmen began to realize how much damage had been done to the Anson's controls. "The use of elevators tended to turn the aeroplane to port while at the same time causing loss of altitude."

In spite of his problems, Grime succeeded in keeping the battered Anson in the air until he put it down safely at Charlottetown. He taxied to the flight line to the considerable interest of onlookers on the ground who saw what appeared to be a sizeable spruce growing horizontally out of the cockpit. After breaking open the distorted hatch with a fire axe, the relieved crew clambered out and surveyed the damage. In addition to the tree in the nose, "an inch was missing from the port propeller, a branch had cut into the port wing between engine and fuselage, exposing fuel lines but fortunately causing no damage to them. The most amazing discovery of all," recounts Robinson, "was that the tubular support for both elevators was cut cleanly underneath the rudder, and both elevator and stabilizer on the port side were missing."

When Avro experts examined the Anson, they declared it unflyable. Robinson is of the opinion that it took "miraculous handling skill" to get the aircraft safely back to base. Undoubtedly the crew was fortunate to have been aboard one of the few original Mark I Ansons left on the base; unlike the later Canadian-built versions, it possessed a metal structure. Robinson is sure that any wooden Anson would have disintegrated the instant it hit the trees.

Robinson was involved in another near thing aboard an Anson at No. 31 GRS. It happened one frigid morning just before dawn when, he recalls, the "weather report predicting fog banks and poor visibility did little to boost morale." The first aircraft started up – reluctantly – after which a figure waving a flashlight came racing across the tarmac, informing everyone that take-off had been delayed due to the fog. "My skipper on this occasion answered to the name of Ted [Haslam] and, although just a youth who, through no fault of his own, lacked experience, was considered by all who knew him to be exceptionally bright and most capable. On this morning, our engines were running smoothly before many other pilots had located their charges." On hearing of the postponement, most of the crews returned to the hangars.

But Haslam decided that the fog was thinning. He gunned the Anson and went trundling along the taxiway in preparation for take-off. The air was perfectly calm. As the Anson arrived at the end of one of the four runways, the duty pilot's Aldis lamp flashed amber.

"A quick run-up of the engines," Robinson says, "and moments later the expected green light as our roll began. Rapidly picking up speed, our attitude changed as the tail came up. Within a few seconds we would attain flying speed. . . . Suddenly the normally

mild-mannered Ted yelled, '*Look out!*' – at the same time ram-
ming throttles through the emergency gate and yanking back on
the control column. An instant later we were almost perpendicu-
lar, hanging on the propellers, seconds away from a stall, when
another Anson flashed past underneath us."

That Haslam managed to wrestle the almost-hovering Anson
back into normal flight testifies to his exceptional skill. But how
could it be that the duty pilot had two aircraft taking off in opposite directions on the same runway? It transpired that there had
been a lengthy delay between the first and second aircraft leaving
the ramp. Thus, says Robinson, "we were completing our run-up
as the second Anson arrived at the opposite end of the runway.
Assuming himself to be the first departure . . . the pilot was not
surprised when given a green light." In fact, he hadn't been given a
green. The duty pilot, unaware that he had two aircraft on opposite ends of the runway, had put his Aldis lamp down, at the same
time pressing the trigger to check on the colour, then, "without
releasing the trigger, sweeping much of the field before focusing
on us, *giving both aircraft take-off clearance.*"

When the BCATP began, the air force faced a shortage of navigation instructors every bit as critical as the shortage of flying
instructors. Fortunately it didn't take as long to train a man to
teach navigation as it did to teach flying. Many ex-high school
teachers found their way into the work. Predictably, they ranged
from excellent to execrable, as they lacked both practical and
operational experience of the work. Ewart Cooper from Winnipeg attended No. 5 AOS in his home town. He remembers one of
the instructors was a former high school teacher "who had taken a
crash course in navigation instruction – but some of the details he
had not yet buttoned down." He was in fact little more expert in
navigation than his pupils, and was learning the subject as rapidly
as he taught it. Many former students remember such instructors

who taught "by rote, without real understanding of their subject," as Frank Phripp puts it. An RAF student navigator, Ron Hunt, recalls that "astronavigation was a bit of a puzzler at first; not so much I think because it was really that difficult, but because the middle-aged non-aircrew instructors liked to believe that they were teaching a very difficult subject."

Few instructors could do much to prepare their students for the realities of the operational world, for only those who had served in combat during World War I had personal experience of battle. None knew what modern aircrews had to contend with; none knew what it was like to solve navigational problems with searchlights stabbing the sky and flak bursting mere yards away, to line up the aircraft for its bomb run while enemy fighters lacerated the darkness with cannon fire and the interior of the aircraft reeked of explosive. Although some tour-expired instructors would eventually find their way into the instructor ranks of the BCATP, the vast majority would always be non-operational.

Budding observers learned to drop bombs and fire machine guns during a six-week course at a bombing and gunnery school (B&GS). The first classes provided about thirteen hours of bombing practice, during which each pupil dropped some forty eleven-pound practice bombs, using the outdated Mark I, II, or III bombsights. Their average error was 274 yards. Within two years, with the advent of better instruction and equipment (the Mark IV Course Setting Bombsight [CSBS] began to appear in the B&GSs), each student was flying approximately twenty-three hours, dropping eighty practice bombs, and recording an average error of 113 yards by day and 136 yards at night.

The student air bomber realized early in his training that accurate bombing depended upon a number of key factors – most of them totally outside his control. He could, for example, put only a limited amount of faith in the winds given to him at briefing.

Unfortunately, the same could be said of many of the staff pilots who flew him to the target. Accuracy in bomb aiming necessitated close teamwork between the air bomber and the pilots, but the training school milieu was hardly conducive to teamwork in the air. Students and staff pilots rarely flew together more than once or twice; they were little more than names in logbooks to one another. Moreover, boredom, inexperience, and ineptitude were present on too many Anson flight decks at B&GS, influencing the pilots' performance in a hundred ways, some subtle, some not so subtle. Add to this the fact that most staff pilots disliked their jobs and could hardly wait to get into more active service and it is hardly surprising that the majority of bomb aimers only realized their potential when they got overseas and became members of operational crews.

When former student pilot John Macfie became an air bomber and went to No. 4 B&GS, Fingal, Ontario, he found that "if you come within fifty yards, you're lucky." The targets were pyramid-shaped wooden structures on oil drums, floating in nearby Lake Erie. The bombs, filled with titanium tetrachloride, emitted a puff of smoke when they burst. Two observers on shore in bomb-plotting towers known as "quadrants" watched the bombing practice and marked the spot where each bomb fell. Upon his return to Fingal, each student was told what score he had achieved. In the rather unlikely event of a direct hit, only minor damage was done to the wooden targets (and the student bomb-aimer who had accomplished the feat had to buy beer for his class). Repair crews usually had to do little more than replace a wooden slat or two. Alan Ramsay, then a flight sergeant senior armament instructor at Fingal, remembers that the repair crews became so casual about the risks of being hit during bombing that they often worked while practice bombs fell around them. As far as is known, no member of a repair crew was ever hit. But other things sometimes went wrong. Mac Reilley recalls a fellow student dropping his eleven-pound practice bombs "on the quadrant hut instead of the target" and catching one unfortunate in the out-house, happily without injuring him. At Trenton, a student observer clambered into the bomb-aiming position of a Battle and set about the task of bringing the bombsight down to its operational position. By mistake, he pulled out a U-shaped pin in the sliding bracket – and the entire sight fell out of the aircraft.

The first bomb-aiming exercises were exciting for students, if not for the bored staff pilots. In the spring of 1943, New Zealander Ron Mayhill climbed aboard Anson No. 8620 at No. 7 B&GS, Paulson, Manitoba, in company with two compatriots. All three of them were agog; it was not only their first bombing "mission" but also their first trip in an aircraft.

> Up we went to three thousand feet, over Dauphin Lake, settling at 120 mph. On the run to the practice bombing range, the pilot waved, and Jim McCormick, who was nearest, crawled over to receive the height, temperature, and airspeed corrections for us to compute on our Form T32s, as we had been trained. Another wave, there being no intercom in the glasshouse cabin, just a speaking tube from the tapered nose to the cockpit, and Jim slithered head first through the narrow aperture below the pilot's feet, into the claustrophobic nose with its small downward window. A few minutes later, a red-faced McCormick squirmed back, the pilot wiping his brow and shouting something which evidently shocked Jim. . . . Around went the Anson in a vicious diving turn that threw Ian Ward and me off our seats. When we had picked ourselves up, Jim leaned over and shouted, "There's no bloody bombsight on board!"

In their excitement, none of the three students had thought to check on the presence of that essential piece of equipment.

The merciless cold of the prairie winter colours many former students' recollections of the BCATP. Ken Roberts of Toronto trained as a bomb aimer at No. 5 B&GS, Dafoe, Saskatchewan, early in 1943. He has vivid memories of the biting chill, which made the job of bombing-up the training aircraft about as appealing as an afternoon with the Spanish Inquisition. The students had to lie on their backs under the Ansons, attaching the eleven-pound practice bombs by forcing metal loops on the bombs into latches on the racks. "We then had to untwist a wire in the nose of the bomb to arm it," says Roberts. "At about twenty degrees below zero Fahrenheit, and with the wind whistling across the prairie, our hands became so cold we could scarcely hold a cigarette lighter. The trouble was that the process was so finicky we couldn't wear gloves." Once in the aircraft and on the way to the target, the tyro air bombers had to set up their bomb sights, which involved "more finicky adjustments with bare hands." Then came the bombing runs, and another ordeal with the elements: "The perspex panels through which we were supposed to watch the target were so scratched and dirty that we couldn't see through them. So we had to slide them out of the way. . . . *Now* we had a hundred-mph blast of frigid prairie wind blowing on our hands and faces."

Wings day for observer/navigators and wireless operator/air gunners was just as important an event as for pilots, and nerves were just as taut. RAF student Hugh Redfern was selected as parade commander for the wings parade at No. 1 CNS, Rivers, Manitoba, early in October 1944. Heavy rain pounded Rivers that day, necessitating an indoor parade. Unfortunately the students "had had only one practice indoors," Redfern recalls. The ceremony had to go ahead anyway. "All went well until the actual march past the station commander, Group Captain Murray, RCAF, after we had received our wings. At the crucial moment, Sergeant Redfern, as parade commander, gave an almighty shout: '*Eyes left!*', when it should have been '*Eyes right!*'" To Redfern's relief, no one obeyed his order; everyone automatically turned toward the CO. Such things happened on important occasions. When Don Macfie (John's brother) graduated from No. 1 B&GS, Jarvis, Ontario, in November 1941, nervous tension caused two of the graduates to take wrong turns during the parade. Macfie himself remembers, "When my turn came, I felt as though there were twenty pounds of lead on each foot."

Wireless operator/air gunners spent twenty weeks – later extended to twenty-eight – at wireless school, learning to send Morse in the smooth style of the professional, unravelling the complexities of QDRs, QEFs, grids and frequencies, D/F loops and R/T. In the early days, there is little doubt that the BCATP failed to train student wireless operators to the required standards. "When we arrived in the U.K.," recalls former wireless operator Cecil Robson, "we found that our wireless standards were not on a par with the RAF. Also, they were using radio equipment in bombers that we did not have in our Canadian training." Fred Chittenden of Hepworth, Ontario, who enlisted in June 1940, says that most of his instructors were "pretty green" and that the speed of Morse code taught in Canada was inadequate for the European war zone; that is, eighteen words per minute. One of the problems was the lack of flying time for student wireless operators, a result of the critical shortage of aircraft and equipment at that period. Nova Scotian Jim Lovelace was one of the wireless operator trainees at St. Hubert, Quebec, in 1940. One day, four of them clambered aboard a Norseman and took off for a hour's trip. This constituted their airborne training. "We had fifteen minutes each on an American Collins wireless transmitter modified for airborne use. It was our task to make contact with base at the wireless school in Montreal. Our ability to use the equipment for communications purposes did not seem to matter. We were considered properly and adequately trained."

BCATP Morse standards soon improved, eventually reaching twenty-two words a minute.

The RAF entered World War II convinced that its bombers could acquit themselves well against enemy fighters, with tight, well-disciplined formations creating massive concentrations of fire-power. The air gunners of the prewar RAF were ground crewmen, armourers in the main, who had volunteered to fly as gunners when needed – for which they earned an extra few pence a day and the right to wear the brass winged bullet on their tunic sleeves. No aircrew badges adorned their chests; the famous half-wing AG brevet had not yet come into being. In fact, the extracurricular gunners did not even rate as bona fide aircrew; they were merely "bods" to be summoned as needed to struggle into flying gear and occupy their turrets, hatches, and, in some cases, open cockpits. After flying, they returned to their regular ground duties. The time spent as gunners did not excuse them from any of their "real" work on the ground. If they left something unfinished when they took to the air, they had to finish it when they returned, no matter what the time. If they missed meals at the Other Ranks' mess, well, it was too bad; they couldn't expect the cooks to prepare meals at inconvenient times of day or night, could they?

When war came, the air marshals found to their dismay that, although the RAF bombers flew in tight formations as they had been ordered, and although the air gunners brought dozens of guns to bear on every target, they seldom hit anything. The reason soon became all too clear: the German fighters carried formidable 20 mm cannons with a far greater range than the .303 Brownings in the British bombers. The Germans soon figured out how limited was the range of the rifle-calibre Brownings. Then they simply stood off at a safe distance and blazed away with their can-nons into the meticulous formations. Scores of Britain's most experienced professional airmen – including dozens of armourers

– were lost on those early air raids which cost so much and achieved so little.

The war was only a few months old when Air Ministry Order (AMO) A552 introduced the air gunners' brevet and decreed that all air gunners would henceforth be sergeants or higher. Indeed, the Air Ministry envisioned commissioned air gunners, although in fewer numbers than commissioned pilots or observers, perhaps a third as many. The original sketch for the AG half-wing sported thirteen feathers. To avoid negative reactions among the superstitious, one feather was eliminated before the badge went into production. [13]

It was at B&GSs that trainee observers and wireless operator/air gunners met. Each had to learn how to handle machine guns, principally the Vickers GO (gas operated) and the Browning, both of .303-inch calibre. Alan Ramsay describes the Vickers gun as a "pathetic weapon with an indifferent rate of fire ... and a magazine that held only sixty rounds." It was already obsolete in Europe. Accuracy in gunnery required skill in the vital subject of aircraft recognition, because knowledge of the wing span and fuselage length of the target aircraft enabled the gunner to use his ring-and-bead sight effectively. At about two hundred yards, says Ramsay, "a Messerschmitt 109 parallel course beam shot would just fill the ring." Every third round was usually tracer, enabling the student gunner to "use the arc of his fire for gauging deflection." Ramsay feels that skeet shooting was "one of the best methods of indoctrinating new gunners in the art of leading the target and calculating in a split second the deflection required." He recalls the famous Canadian fighter pilot Buzz Beurling visiting Fingal and telling students how he learned about deflection shooting from duck hunting as a boy.

The BCATP pioneered several classroom teaching methods. A contemporary U.S. article reported: "Even classroom work is lively, thanks to the Canadians' resourcefulness in what is known as synthetic training. In one room students peer through regulation gun sights at moving scale models of Heinkels and Junkers and Messerschmitts, etc., which they are expected to identify instantly with an estimate of range. In another room is a full-scale rotating aircraft turret with a gun that aims a beam of light at a moving model of a Messerschmitt 110. When the light hits a photoelectric cell in the fighter's nose, a meter registers a hit." [14]

The extent of gunnery instruction varied widely, particularly in the early days, since it depended on the availability of aircraft, equipment, pilots, and instructors. Air gunners were lookouts as much as defenders of heavy aircraft. Some flew entire tours without firing a shot in anger, but their sharp eyes kept their aircraft out of trouble. At B&GSs, student air gunners learned to deliver their "running commentaries" by which they kept their skippers informed of what was happening to the rear, side, or below. They mastered the mysteries of turret manipulation, sighting, estimating range of targets, gun stripping, dealing with stoppages, harmonization of guns – and, of course, aircraft recognition.

Accommodation for student gunners in the Battles was hardly luxurious. John Wullum, who joined the air force in June 1941 and trained at No. 9 B&GS, Mont-Joli, Quebec, remembers the "noxious carbon monoxide and glycol fumes" that swept down the fuselage to where the students gunners were positioned. In the not-infrequent event of a jam, the tyro gunners had to strip and reassemble the Vickers gun on the floor of the aircraft. Occasionally the staff pilots got bored while the stripping and reassembling was going on and indulged in a few steep turns and dives. The result, says Wullum, was "a lot of small pieces . . . floating around in the fuselage." During his training in Canada, Wullum had only 2.55 hours on power-operated turrets. In 1942 some Battles were fitted with turrets in place of the Great War-style open gunner's position. At about the same time, Bolingbrokes came into widespread use at gunnery schools, with power-operated gun turrets as standard equipment.

Fledgling gunners practised their skill on twenty-foot sleeve targets dragged around the sky by Battles and Lysanders. By early 1942, when Alfred Tait was training at No. 4 B&GS, Fingal, Ontario, the Battles were in sad condition. He recalls going out on exercises again and again, only to return because of mechanical problems with the aircraft. Three Battles crash-landed one day due to landing-gear problems. More serious accidents sometimes befell target towers. Alan Ramsay remembers a Lysander at the same school coming in to drop a drogue before landing. The pilot misjudged his height, and the drogue cable snagged a tree. In most cases the rapidly moving aircraft would have ripped the cable free. In this case it didn't. The cable held. With its engine bellowing helplessly, robbed of the speed which kept it aloft, the Lysander stalled and fell out of the air like so much scrap metal. The drogue operator in the back seat jumped without a parachute. He landed flat on his back and was killed instantly. The pilot, who remained in the aircraft, also died.

Jim Lovelace was one of the first wireless operator/air gunners to go through the BCATP. He recalls that much of his time at No. 1 B&GS, Jarvis, Ontario, was spent at ground school, where he studied the workings of the outdated Vickers gun and did target practice. "After mastering the vocabulary of 'rear sear release' and the 'rear sear release levers,' 'rear sear release lever covers,' together with the 'rear sear release cover pins' and the most vital but tiniest parts of all, those 'rear sear release cover pin retainers,' and God knows what else, we were considered expert enough to get behind one in an aircraft," he says. But first the budding gunners had to be outfitted for their aerial duties "in a kind of canvas coverall with a weird arrangement involving one leg like an ordinary pant leg, but the other zippered from the ankle to the neck." Lovelace and his fellow students had been briefed to urinate before kitting up, as it would be impossible later:

There was no way one could extract oneself from that stranglehold of parachute harness, flying suit, heavy-duty six-button fly and long johns . . . ! A final addition to this outfit that resembled nothing more than a deep sea diver's suit was a World War I flying helmet and Gosport tube. The latter looked like a doctor's stethoscope at one end and a syphoning tube at the other. "Never mind," they told us, "you'll get to know how to use it when you see the aircraft connections inside!"

Bundled up, we had to trundle across the hangar to get our parachutes. From the parachute section we made our way to the armouries in another hangar to get our panniers: flat cylinders of sixty rounds each, .303 ammunition. We were shown during one class how the armourers loaded these panniers. Each cartridge was separately inserted into an opening and pushed into the cylinder. When sixty rounds had been inserted, a handle on the top was wound like a clock to provide tension to release the cartridges as the machine gun was activated. This handle was supposed to be folded into a recess and locked. When needed, it could be unfolded and used as a carrying handle. They must have run short in the RAF before we got them, because there was only one handle in the armouries. It was always a balancing act from the armouries to the aircraft, all revved up and waiting. Sixty pounds of pannier, a parachute and loose leggings with a floppy helmet and wildly swinging Gosport between one's legs, all caught up in whirling prop wash. The pilots always waited until we put a leg up on a slippery aluminum retractable spiked step to throttle the engine, providing some comedy relief. . . . It was pure hell for the trainee air gunner. . . . He was anchored to the oil-slick floor with a heavy webbing static cord clip-fastened to a crotch-ring on his parachute harness. At least twice during each pass, when the drogue would be just about

perfectly lined up in the ring-and-bead sight, the pilot
would pull a turn that either lifted one bodily three feet
off the floor or shoved you to your knees.

Howard Hewer, whose ITS course was posted *en masse* to wireless school, went through gunnery training at No. 1 B&GS, Jarvis.
He has vivid memories of aerial gunnery practice. On January 28,
1941, he clambered aboard a Battle with Frank Cook, a friend from
Toronto's Parkdale Collegiate. Scheduled for a camera gun exercise, the two students immediately noticed that the wire, or
G-string, which anchored them to the floor of the gunner's
position, was missing. Inexperienced and not wishing to cause
problems, they said nothing about it. Soon they were flying at six
thousand feet over Lake Erie. Hewer recalls: "I was happily firing
away with the camera gun at another Battle when the pilot pulled
up the nose sharply – and over the side I went!"
    One of Hewer's heels had jammed in the metal gun-ring, while
the rest of him dangled in the frigid slipstream, still clutching the
camera gun in one hand. "I had time to look down at the ice in the
lake below, in some horror, before Frank was able to reach over
and grab the seat of my pants."
    But while Cook attempted to drag Hewer back into the Battle,
the pilot, blissfully unaware of the drama going on behind him,
"was happily swooping and turning all over the sky." Hewer
remembers shrieking at Cook, "For God's sake, call him to level
out!" Cook didn't obey. To reach the Gosport tube, he would have
had to relinquish his grip on Hewer's posterior. Besides, it was
questionable whether the pilot would have understood the message. In the Battle, the air pounding about in the gunner's position
usually made conversation between the cockpits little better than
"a moose moan," as Jim Lovelace describes it. Fortunately the
muscular Cook was eventually able to haul his friend back into
the aircraft. On landing, the two students made no mention of the
incident, although, Hewer notes, "we did report the absence of the
G-string."

A student gunner's first air exercises were simple: merely a question of flying alongside the drogue and learning to "lead" to compensate for the relative speeds of target and gunner. The drogue flew puffed-out and firm like a windsock in a gale. It appeared to be an easy target. It wasn't. Vast quantities of ammunition streaked around the sky until the student gunners learned to compensate for the shifting and skidding, the swooping and soaring of their targets. Every student gunner used his own ammunition, the rounds having been dipped in blue or red wax. After target practice, ground staff examined the hits on the drogue, checking on the smudges of colour and counting each man's hits. Student air gunners were frequently dismayed at how few hits they scored. Don Macfie was jubilant when he scored fifty hits out of four hundred fired. "Just luck," he admitted to his diary.

Airsickness plagued many wireless operator/air gunners, just as it did other aircrew. Frank Murphy of Toronto was "violently sick" every time he flew at No. 1 B&GS. "I reported my sickness to the MO," he says, "and he told me that if I got sick again, he would arrange to give me a ground job. There was no way I was going to lose my chance to be aircrew, so I never again reported that I was airsick, even though every time I flew in the RCAF I was violently ill." George Webb of Hamilton was similarly afflicted. Like Frank Murphy – and countless others – he continued to fly, simply putting up with the unpleasantness. Webb regularly vomited into his nylon glove liners, throwing them away at the first opportunity. But he had another problem: "On occasion I would lose my upper bridge. Finally the station dentist got wind of this and it was either remove the bridge before flying or not fly at all!"

When France fell, in June 1940, the British high commissioner to Canada, Sir Gerald Campbell, informed the Canadian government that Britain was anxious to move four service flying training schools out of the war zone to the safety of Canada. Ottawa didn't object; the schools could be accommodated in Canada without creating any problems for the BCATP, provided the British agreed to be responsible for the operating costs and part of the capital expenditure on new airfields.

On a heavily overcast afternoon that same month, a Lockheed Hudson of the RCAF suddenly appeared through heavy clouds near the town of Bowmanville, about forty miles east of Toronto, on the shore of Lake Ontario. Some eye-witnesses claimed the twin-engined bomber was already burning when it came into view; others said it burst into flames after turning on its back and diving vertically into the ground. The aircrew died instantly, so did their passenger, Norman Rogers, Mackenzie King's minister of national defence. Rogers had been travelling to Toronto to deliver a speech. He had left Ottawa shortly after noon; the regular TCA flight, on which he had intended to make the trip, had been cancelled because of bad weather. To replace Rogers, King appointed his finance minister, Nova Scotia-born James Layton Ralston. A much-decorated battalion commander in the Great War, Ralston had a deep-seated suspicion of men with blue eyes. Some claimed it was because of his experiences with blue-eyed Mackenzie King. The meticulous Ralston proved to be a highly successful choice for this sensitive post, although he and King would eventually fall out over the conscription question.

During that eventful period another Ottawa appointment occurred, and it had considerable impact on the BCATP. Two hopelessly disparate personalities had collided head-on when the recently appointed minister of national defence for air, "Chubby" Power, met the efficient but somewhat straitlaced chief of the air staff, Air Vice-Marshal George M. Croil. Although Power considered Croil a competent, hard-working officer, he found him too "regimental" and got the impression, "rightly or wrongly, that friendly, sympathetic cooperation with him would, owing to our fundamental differences of temperament, be difficult if not impossible. I wanted friendship and cooperation; he, I imagine, expected me to give little more than routine supervision, leaving to him the unquestioned authority over the members of the service, and possibly over the purely civilian functions of the department."[15] Croil was shunted off to the sidelines, becoming inspector general of the RCAF, a post specially created for him and carrying prestige but little power. His successor was that Falstaff-in-blue, Lloyd S. Breadner, who was as fond of a drink or two as was Power himself. Predictably, the two got along well.

Later that year, Ralston and Breadner travelled to England to attend a conference on the future of the BCATP. Now that France had fallen and Italy had joined Germany, the British Commonwealth confronted the Axis powers alone. Ralston flew in an unheated bomber and caught a chill which resulted in a bout of sciatica. He had to spend most of his time in Britain confined to a wheelchair. The minister of munitions and supply, C. D. Howe, nearly lost his life getting to the same conference when a U-boat torpedoed his ship, *Western Prince,* south of Iceland. Howe was fortunate to survive the ordeal, which included several hours in a lifeboat.

For the Canadians who had gone through so much to get to the conference, the results were disappointing. They soon discovered that the BCATP, the major focus of Canada's war effort, counted for little in a London where survival itself was still in question. Archibald Sinclair, the urbane British secretary of state for air, informed the visitors candidly that the Air Ministry was far too busy coping with German air raids to concern itself about the BCATP. In effect, he told the Canadians to keep on with what they were already doing. Moreover, Sinclair urged them to proceed as rapidly as possible with the local production of twin-engined Anson trainers, because there was no telling when the type could be released for export from Britain again. Scores of Ansons were now being held in Britain for use in the event of an invasion.

This was not good news for the Canadians. Britain had so far delivered less than sixty Ansons. Although the idea of building the aircraft in Canada had sounded good to everyone, the unpleasant truth was that the Canadian aircraft industry, which had been allowed to shrivel almost to extinction between the wars, was having trouble handling the job. There simply weren't enough aircraft engineers and experienced manufacturing personnel available. By now the shortage of Ansons was affecting training schedules; for example, No. 7 SFTS, Fort Macleod, Alberta, was struggling along with only three of its scheduled complement of thirty-six Ansons. When RAF trainee Douglas Wadham arrived at No. 33 SFTS, Carberry, Manitoba, in February 1941, he expected to be flying Ansons, but the shortage necessitated a switch to single-engined Harvards. Like many trainees he found the powerful Harvard a handful to control, noting in his diary that he was "doubtful about my capabilities with this machine." (His capabilities proved adequate, however; he won his wings and eventually became the captain of a Catalina flying boat.)

The disasters in Europe had already had an impact on the training schools in Canada. The need for aircrew had become critical. The BCATP rose to the challenge, accelerating the whole plan, opening eight instead of two EFTSs by mid-1940. It was a brave move, for no one was sure that the advanced schools, the SFTSs, would be ready in time to accept the increased numbers of pupils. The new plan called for eight instead of five SFTSs by year's-end; indeed, the speed-up order applied to AOSs, B&GSs, et al. The successful acceleration of construction schedules was an astonishing achievement, one that probably did much to convince the British that the Commonwealth Air Training Plan, on which so much depended, was in remarkably capable hands.

In September, the first British advanced flying school was established in Canada: No. 31 SFTS, at Kingston, Ontario. Originally scheduled for completion in June 1941 to train Fleet Air Arm pilots, the facilities were in operation by October 1940, before any

FAA students had arrived. Two courses of non-naval students graduated, after which the school reverted to its original purpose. Toward the end of 1940, four more RAF schools moved to Canada: No. 32 SFTS to Moose Jaw, Saskatchewan, No. 33 SFTS to Carberry, Manitoba, No. 31 ANS to Port Albert, Ontario, and No. 31 GRS to Charlottetown, Prince Edward Island.

The British schools were not part of the BCATP, although they were in most respects almost identical to their Canadian counterparts. They operated under the Visiting Forces Act, which enabled them to keep their national identity while operating under the control of the RCAF.

CHAPTER SIX

Town and Country

Cessna Crane

"The English are not renowned for wearing their hearts on their sleeves, but I would not want to leave this world without saying, 'Thank you, Canada, for many happy memories, for some good friends and, above all, for your unstinted help from 1939-45.'"

– former BCATP student and instructor Jeremy Howard-Williams, Southampton, England

For dozens of communities across Canada, the creation of the BCATP was the best news they had heard in ten years. Once the plan got going, they all said, it would be a veritable bonanza. First, scores, possibly hundreds, of construction workers would arrive. Burly, free-spending men with healthy appetites. And they'd only be the beginning. Next would come the air base staff, accompanied by wives and children in many cases. A marvellous boost for the local economy, which, goodness knows, had been virtually moribund for as long as anyone could remember. Then, finally, the airmen themselves, the students, splendid young fellows, the cream of the crop, a lot of them from monied families; they'd have cash in their pockets and the inclination to spend.

In Yorkton, Saskatchewan, the local paper could barely contain its enthusiasm about the establishment of a training school in the vicinity. "It will bring about 1,000 men here with an average pay of about $100 a month.... This will help all lines of business in Yorkton. Airmen have clothing to buy, suits to be cleaned, shoes to be purchased, yes, and shoes to shine. They will frequent restaurants and hotels, and in a nut-shell will leave $100,000 a month with Yorkton business institutions.... Yorkton will be in for one of the greatest booms in its long history." [1] But, the newspaper cautioned a couple of weeks later, there was always the grave possibility of an early peace breaking out, in which case there would inevitably be "a reaction and all business would suffer. It might be that the board of trade and city council should get together and consider the advisability of restricting, through license, too big a rush of business entries to the detriment of established business." [2] The paper evidently expected gold-rush conditions to prevail in the near future, with all manner of dark implications for the town: "We should take steps to keep undesirables out of the city. This ilk follows the opening of camps where large bodies of men are congregated, like leeches; they seek to profit" (unlike the local merchants, apparently). "These kinds of people have no scruples about taking advantage of the boys who are serving their country. By careful scrutiny our police force might do considerable [sic] along these lines." [3]

Yorkton's sentiments were typical of those of dozens of small towns across the country. Every one of them wanted a BCATP base or depot or, more precisely, wanted the economic benefits that such units generated. Even back in 1929, when the town of Trenton, Ontario, was selected as the site of the RCAF's new training centre, the economic implications transported the local business community into raptures. "The stupendousness of the announcement is hard to realize," chortled the *Trenton Courier-Advocate.* "The amount of money involved is enormous. The benefit to Trenton can be imagined, but no one can anticipate to the fullest this wonderful news at the present time."[4] Ten years later, the prospect of scores of BCATP bases being built across the country set civic pulses pounding anew. Dozens of delegations from towns large and small hurried to Ottawa, maps and statistics at the ready, eager to convince anyone who would listen that they had the right spot for a training base. In Moose Jaw, Saskatchewan, which then had a population of just over twenty thousand, the mayor was crestfallen to learn that the city's name was conspicuously absent from the first list of communities issued by the BCATP authorities. He lost no time in wiring the minister of national defence: "Moose Jaw City Council pledges fullest cooperation of Moose Jaw citizens in Canada's war effort. . . . People disappointed that no use being made of superior natural advantages of terrain and visibility of City and district in Commonwealth Air Training Program. . . . If Moose Jaw is to be included, suggest the importance of immediate notification to permit planning for accommodation, housing, etc. Moose Jaw Armoury little used since outbreak of war."[5] The mayor needn't have worried. Moose Jaw had already been chosen as the site for a training establishment to be built for the RAF, but for political reasons – 1940 was an election year – no announcement had yet been made. (As we shall see, the mayor may well have wished that his city had not been selected.)

If the local politicians in those prairie towns seemed to be overreacting, no one could blame them. The past ten years had been the worst in anyone's memory, bringing a numbing succession of disasters both economic and natural. No one could remember such wind and hail storms, such droughts, such plagues of grasshoppers, such blinding blizzards of dust. "There was no escape from the dust," wrote Pierre Berton of that period,

" – the insidious, all-pervading dust – that crept ghostlike into the tightest houses, seeping through cracks in doors or window frames or down chimneys until everything – furniture, carpets, bedspreads, window sills – was covered in a spectral mantle of grey powder." During the worst of the droughts, nature mocked the desperate farmers by producing catastrophic barrages of hail instead of sorely needed rain. "Thousands of acres of wheat were shredded. Granaries were blown down, barns wrecked, horses killed. Thousands of turkeys, chickens, waterfowl, and rabbits were battered to death. Windows were shattered and telephone lines crushed as the winds whipped up to one hundred miles an hour."[6] Families by the score packed up and moved away (*somewhere* there had to be better conditions), leaving their homes to stand as melancholy memorials to better times. But moving often created even more problems for such families, for they found that by migrating to a new province they had rendered themselves ineligible for relief. And they lacked the means to return home, even if there had been anything to go home to. The result? More suicides. More misery. The economic system seemed to be coming apart at the seams. With a catastrophic tumbling of local tax revenues, municipalities found themselves unable to pay policemen, engineers, teachers, or any of the other cornerstones of civilized communities. Wheat prices tumbled to a quarter of their 1929 levels. Many Canadians starved to death. The sheer humiliation of being on relief killed others.

Work went on at the Trenton base throughout the worst depression years, because it was new and qualified as a relief project. No doubt the labourers and craftsmen counted themselves fortunate to be working at Trenton; any job was better than none. Unmarried, unskilled men got their board, food (standard military rations at a cost to the government of twenty-six cents per day per man), and clothing, plus five dollars a month – or about twenty cents a day – to spend as they liked. According to the press and self-interested camp officials, the "twenty centers" lived lives full of jolly comradeship and healthy activity. In fact, the workers' lot was far less congenial than was generally realized. Once they signed on, they became virtually indentured. True, they had the right to quit if they so decided. But quitting resulted in their names vanishing from the relief rolls – the authorities' favourite way of keeping the dissatisfied in line. In Trenton the threat was a powerful incentive to stay and put up with the spartan living conditions – in most cases tarpaper shacks and primitive sanitary facilities – and the niggardly pay.

Few Canadians complained when prices rose after the war started. The employment situation was looking up; in fact, workers were soon in short supply. At Yorkton, the new $800,000 BCATP aerodrome project provided work for most of the town's unemployed. By September 1941, the Relief Committee noted that "although the cost of living had 'increased sharply,' the actual cost of relief in the city had 'dropped greatly.'"[7] At the peak of the construction activity which produced No. 11 SFTS, some five hundred men were busy on the site.

It is interesting to compare the experience of Yorkton with Weyburn, another Saskatchewan town of similar size. The war had been on for nearly two years when Weyburn learned that an RAF school (No. 41 SFTS) was to be established there. The news sparked no euphoria like Yorkton's. Already accustomed to wartime prosperity, Weyburn reacted with quiet satisfaction. The airfield would be a good thing for the local economy, noted the local newspaper, but perhaps its greatest benefits were yet to be realized. An editorial in the *Weyburn Review* pointed out that "the airport will be there when the war ends. . . . Coast to coast air transportation will be in for a boost requiring feeder lines to supply it with business, both passenger and freight, and for a community to be without an airport will be about as bad as being without a railroad."[8] Such was progress. A few years before, mere survival was an achievement; now the town serenely contemplated a bright future as an airline centre.

The construction of No. 41 SFTS involved a number of outside contractors, but much of the work was handled by local labour. Construction at the air base proved to stimulate business in general, prompting the asphalting of several blocks of the downtown area by a company engaged in similar work at the airfield. The impending influx of air force and other government personnel stirred the local authorities into doing something that had never before seemed necessary: putting nameplates on every street and avenue intersection, more than a hundred in total. The good times enjoyed by Weyburn, Yorkton, Dunnville, Jarvis, and scores of other small communities across the country, were in large measure due to the construction of BCATP bases. As historians Brereton Greenhous and Norman Hillmer remarked in their study of the impact of the BCATP on some Saskatchewan small towns:

> Businesses revived or were begun because of the BCATP; men and women found work building or maintaining or serving the base. Bus and taxi firms, drug stores, shoe repair shops, restaurants, beer parlours, movie houses, hotels, dance halls, clothing stores, laundries, barber shops and even churches flourished more than might be expected in relation to a generally buoyant economy. In Yorkton, Lawrence Ball found the taxi service "in terrible disrepute" in 1939. He bought four new Dodges, he recalls, the basis of "a nice business." The transportation entrepreneur in Weyburn was Norman Buss, who improvised a service between town and the base by using a truck which he covered and provided with benches along the sides. Connor Bus Lines later took over the business when the truck was declared unsafe.[9]

The small rural town of Mossbank, Saskatchewan, had a prewar population of about six hundred. In 1940 construction began on No. 2 B&GS, and within months the population had more than tripled. Local resident John Inglis remembers that Mossbank quickly became "a boom town." But good times brought their own problems; the local Chinese Cafe had to cut back on services because the proprietor couldn't hire help, and the blinds came down at the Mossbank Hotel when the staff closed up early, weary from overwork. Finding accommodation became a major challenge. "Every place in Mossbank was soon occupied.... Some lived in garages and some even used tents in the summertime. Small houses of less than 1,000 square feet had as many as three families living in them."[10] Finding somewhere to live became difficult in Weyburn too. "The shortage of small, modern houses for residential purposes has been acute for some time past, but with an expected further influx of people into Weyburn due to the establishment of an air training centre here the shortage of houses will present a major problem."[11] Hundreds more houses were needed in Weyburn without delay. "At the present time even the cabins at the tourist camp are being leased by the month to provide temporary living quarters for federal government engineers engaged on the new airport."[12]

Yorkton experienced a huge building boom due almost entirely to the construction of the local air force base. In 1941 building permits to a value of $200,000 were issued in the city – close to a seven-fold increase over 1940. But all the new building didn't solve the town's housing problems. In the spring of 1942, a garage owner obtained permission to convert a "large portion of his new garage on First Avenue" into apartments. The *Enterprise* commented that this would be welcome news for those "forced to live in basements and other unhealthy makeshift quarters." Equally welcome was the news that the Yorkton Board of Trade planned to utilize the cabins of its "Auto Camp" (an early form of motel) as "permanent" accommodation; soon seventeen cabins, now fully winterized, housed fifty-eight members of the air force. By October 1942, at least three hundred air force families resided in the city, according to the *Enterprise*. The paper calculated that the population had risen by 1,500 since the last census, which had recorded a population of less than six thousand. [13]

In Moose Jaw, about a hundred miles southwest of Yorkton, the lack of accommodation for a burgeoning wartime population was causing concern even before No. 32 SFTS opened there in December 1940. The War Services Auxiliary Council reported that only a very limited amount of rental space had so far been made available for the airmen and their families: single bedrooms at twelve dollars per month, bedrooms plus sitting rooms at fifteen to twenty dollars per month, and furnished suites at fifty dollars per month. Estevan, in the southern part of Saskatchewan, close to the U.S. border, experienced similar problems. In September 1942, city authorities requested Ottawa's permission – necessary in wartime because of shortages of materials and labour – to build a small number of houses to help relieve a housing crisis brought on by the establishment of No. 38 SFTS five months earlier.

Housing problems persisted throughout the war wherever BCATP bases were located. At Hagersville, Ontario, instructor John Clinton was thrilled to find an apartment for himself and his wife – and ecstatic that it possessed indoor plumbing, for all too many renters had to put up with outhouses. In January 1944, when the base at Weyburn was transferred from the RAF (No. 41 SFTS) to the RCAF (No. 8 SFTS, formerly based at Moncton, New Brunswick), the incoming air force families found themselves living in winterized tourist cabins or in rented houses divided and subdivided to create the maximum number of rooms. Property owners generally did well. When No. 31 ANS opened in November 1940 at Port Albert, Ontario, near London, "the whole surrounding community profited. . . . Many residents and farmers . . . in the vicinity took in boarders." [14]

The busier the bases, the more modestly proportioned the available accommodations. Flight Lieutenant Dave Campbell, of the Central Navigation School (CNS), Rivers, Manitoba, had strong feelings on the subject:

> The average size of a one-family dwelling ... is about the same as two fair sized closets attached to a mail box. It used to be considered unsporting to keep anything larger than a pekinese dog in a space that size; but with a little training you'll get used to it. . . . In the winter you'll find that you have to open both the door and the window in order to get room to put on your greatcoat. . . . Some of the obvious expedients like using nothing but condensed milk, petit fours and dwarf peas will readily occur to you. The finer points involved in living like a turtle, however, should be picked up as soon as possible from some of the old-timers. You can readily distinguish an "old-timer" by asking him to open a bureau drawer. If he carefully pulls in his stomach before pulling out a drawer he has been at it for more than three months. [15]

Veterans of renting became inured to the tyrannies of landlords and landladies. Alexander Velleman was a ground crew NCO whose wife journeyed with him to his various BCATP postings. At Portage la Prairie, the couple rented a second-floor room from a Mrs. Rand. Velleman himself had to sleep at the base most nights, but when he had an opportunity to join his wife, he took it. Then to his dismay he discovered that on every occasion he slept "at home," Mrs. Rand was charging him "an extra dollar over and above the agreed amount." He complained and was told that "this was the way it was and it was not negotiable." [16] But not all landladies took advantage of the times. While training at No. 6 SFTS, Tom Anderson found excellent accommodation for his wife at the home of Mrs. Werner on Broad Street in Dunnville, where she was treated "like a princess." [17]

What of relations between citizens and air force personnel? Many ex-airmen who trained in the BCATP have fewer memories of the people living near the bases than might be supposed. The reason was simply that training took up all their time. Nova Scotian Arthur Bishop, who trained at No. 3 EFTS and No. 6 SFTS, recalls: "The pressure of EFTS and SFTS was just too heavy for much fraternizing. . . . Whatever interchange there was was positive, and I can only remember the many kindnesses shown to me and my friends. I will always have fond memories of London and Dunnville." Although students may have spent little time in neighbouring communities, permanent staff were another matter. Instructors, mechanics, clerks, cooks, and every other type of administrator and tradesman became familiar figures in the towns, many living with their families in the communities. In general, relations between the air force and the local citizenry were excellent; inevitably problems occurred, but they were the exception rather than the rule.

In most communities, citizens had been well prepared for the tidal wave of blue uniforms. In Yorkton, for example, the local paper had advised citizens that they "should consider beforehand the certain responsibilities which will be ours in relation to the men of the air force who will make this their recreation centre every day of the year. In the first place there will have to be cohesion of effort in planning suitable entertainment for the men. . . . We are liable to find five or six hundred men of the school in the city every night on leave. We probably should have suitable lounging quarters for those who so desire, with facilities for letter writing, etc. . . . And one thing we should be prepared to do is extend every courtesy and hospitality to the boys here in training." [18]

The editorial writer reasoned that such hospitality might be of long-term benefit for the town. "Who knows but that by these associations and the contacts made we will have a number of them come back to live here after the war. There are many worse little cities to live in than Yorkton." [19]

Yorkton did its duty by the newcomers. When No. 11 SFTS moved in, the *Enterprise* produced a special supplement explaining precisely what a service flying training school was and what it did, as well as describing the organization and operation of the base. A local ladies' organization entertained fifty-seven wives of officers and other ranks at a tea party; a week later the city organized a banquet in honour of the base and its occupants. Hostess clubs provided entertainment for air force personnel, with accommodations supplied by the city; service clubs and citizens donated the equipment and furnishings.

After the shortages of blacked-out Britain, RAF students found Canada a glittering haven of plenty. When Douglas Wadham from Gateshead, County Durham, arrived in Winnipeg early in 1941, he "just wandered around, enjoying the novelty of lighted streets and shop windows." Ernie Allen, a Londoner, still remembers the steak and the banana split he relished on his arrival at Moncton. "This was really living," he says, "after two years of rationing in Britain." Jock Lovell from Blackpool had sailed into Halifax on December 29, 1940, his kit bag bulging with wildly inappropriate tropical gear: "We were entranced . . . the country was lit up like a Christmas tree . . . to be in a dockyard that was just a blaze of light was something that had every man on deck, just looking." Later, on the train to Toronto, he revelled in the excellent food, delighting in the fact that there were "unlimited amounts of it, just for the asking." Douglas Wadham was similarly impressed by the railway meals, industriously noting down the details of every one on the long trip west to Carberry, Manitoba. "Supper was so good, we just rolled around in ecstasy after it. The sweet [dessert] was apple pie – a marvellous dish with firm sweet apples and a creamy crust. . . . It is like a tremendous school treat on this train and we keep pinching ourselves to make sure that we are awake." The next day, the treat continued: "Our breakfast was superlative: creamy soft porridge, sweet coffee, more nice bread, sausages and bacon. I'm beginning to think we don't know how to cook in England – even in peacetime!" LAC Phillip Del Rosso, another Londoner, also became a devotee of Canadian railway cooking after a few mouthfuls, describing it in a letter home as "incredible." "You can have as much sugar as you like [sugar was strictly rationed in England at the time] and plenty of eggs, bacon, jam, oranges, apples, bananas, milk chocolate and sweets." Scotsman Bob McBey gazed in awe at the selection of candies and chocolates. "You mean you can just *buy* it?" he queried, accustomed to having to produce ration cards for such delights – if they could be found.

Hugh Redfern found Canadians "generous to a fault"; John Saqui remembers many Canadians stopping him in the street and asking him home for a meal. "The Canadian people are so friendly," wrote Phillip Del Rosso to his mother, telling her about his nights on the town, courtesy of generous Calgarians. He added: "The girls here are very modern-minded. . . . They have bags of chaps, and practically as soon as you meet them, they take you home to meet their parents, where you are given coffee and cookies." Douglas Wadham encountered a man in the Winnipeg Art Gallery who unhesitatingly invited him to spend the weekend at his home. Wadham noted with interest the differences between British and Canadian homes, paying particular attention to the windows, which "had double frames, with the outer frames replaced by anti-insect mesh in the summer." Jim Smythe, a Fleet Air Arm student pilot at No. 12 EFTS, remembers that the citizens of Goderich, Ontario, acted as if they had "never seen a British sailor before, so they welcomed us with open arms."

British aviation historian Roger Freeman writes: "In general, the Canadians treated the British trainees royally and endured with understanding the occasional incident that put them and their property at risk." He cites one such case:

D. A. Reid was privy to one mishap that could have had nasty consequences: "In 1944 I was sent to Canada for training, taking a course as a Bomb Aimer at No. 5 Bombing and Gunnery School, Dafoe, Saskatchewan. The usual bombing exercise had pilot plus two u/t (under training) bomb-aimers with a load of 12 practice bombs in an Anson. The bombs had a small charge, sufficient to make a bang, flash and smoke that would mark the point of impact for the aimer some 10,000 feet above. Each trainee dropped six bombs singly, taking turns. One night, as the Anson banked over Dafoe town to head for the target, I looked down in the nose and saw my mate, Ginger, drop the release 'tit' on the floor. The bomb-released light went out as a practice bomb descended on to Dafoe town. We abandoned the exercises and returned to base; Ginger was a very worried man. Next morning a townsman arrived at the main gate and was taken to the CO. Presently Ginger was called to report to the Office. We all expected the worst. A little later Ginger came back grinning. The bomb had detonated just outside the bedroom window of the Dafoe man's wife. She had been in labour at the time with a doctor present. The doctor said it was the quickest delivery he'd known!"[20]

In Winnipeg, welcomes for BCATP airmen were consistently lavish and well organized, despite the fact that most of the beneficiaries spent only a few hours in the city. Winnipeg's location made it a natural stopping-off point for the trains carrying air force personnel across the country to various training bases. Bleary-eyed cadets from England, Australia, New Zealand, and elsewhere, young men who had been travelling for several days through unfamiliar territory, imprisoned in railway cars in unwelcome proximity to far too many of their sour-smelling fellows, were totally unprepared for the welcome they received in Winnipeg. Mouths dropped open in amazement as hostesses greeted them with coffee and cookies, no matter what the time of day or night. Countless ex-airmen remember dancing with pretty girls on a platform at Winnipeg in the bone-numbing early hours of winter days. Many must have wondered if they were experiencing some realistic hallucination brought on by their endless journey. If the stopovers were long enough, there would be invitations to join Winnipeg families for a meal or to visit a hostess club set up in a nearby office building, complete with easy chairs, plenty of magazines, and unlimited refreshments. Tet Walston, from Morley, near Leeds, recalls the long trip from Halifax to Calgary: "Whenever the train stopped at a town, we were almost overwhelmed with gifts of goodies from the local citizens." But the highlight of the journey was unquestionably Winnipeg, where music welcomed the airmen. "An RCAF band was playing 'Stardust' as we poured onto the concourse."

Richard Taylor who had travelled from Reading, England, remembers "a huge welcoming group" at Winnipeg, where "addresses were exchanged and many friendships started." Douglas Wadham's group "marched jauntily through some swing doors and turned left to find ourselves alongside one wall of a large hall . . . [and] a crowd of Canadian girls and a few men clapping and cheering. . . . Our amazement changed to confusion and embarrassment as we heard a band playing us in with some stirring march. I was as red as a turkey cock and we were all marching along with eyes on the floor, giving occasional quick glances at these wonders around us. At the end of the wall we right-wheeled up a few steps onto the raised portion, where we were each handed a packet of chocolates and cigarettes. We then filed past tables, behind which stood dozens of fair females of all ages, smiling, shaking hands with us, and handing us whopping big juicy red apples." The airmen, says Wadham, were "overwhelmed by this marvellous display of true hospitality."

Some chance meetings at the CPR station at Winnipeg developed into more significant relationships. During one stopover on his way to No. 32 EFTS, Bowden, Alberta, RAF student Bob McBey met June Turner, a volunteer with the Winnipeg Women's Air Force Auxiliary. When he graduated, the couple became engaged. They married shortly after the war when McBey returned from overseas duty.

The Winnipeg lady volunteers and the air force veterans of the Great War set about their welcoming chores tirelessly and with admirable efficiency, a fact that encouraged the *Free Press* to indulge in suitably warlike analogies:

Reception arrangements for overseas airmen in Winnipeg might be likened to the streamlined organization on an operational fighter station. There, pilots are at readiness as the reports of hostile aircraft filter in to the operations room. Scramble is the order and pilots are off to intercept enemy planes. Always, the ladies of the Winnipeg Women's Air Force Auxiliary and the men of the Wartime Pilots and Observers Association are at readiness; No. 2 Training Command auxiliary services might be playing the role of operations room; official reports of the arrival of overseas airmen converge on No. 2 Training Command from the eastern and western ports. There's the order to scramble and the boys in blue are intercepted under pleasant auspices at Winnipeg train depots. Long trains, each week, spill out Belgians, Czechs, Poles, Free French, English, Scotch, Australians, New Zealanders, or Norwegians for a breathless welcome at Winnipeg before they reboard for dusty flying fields in the east or west. Like bright flowers on a navy blue table, flags brought by the consuls of different nations light up the station rotunda. Airmen swarm to the friendly flags of their homelands. The RCAF band strikes up Colonel Bogey, there are heaps of cookies and gallons of tea, pretty girls to chat with, charming hostesses to say a kindly word, veteran fliers of World War I in mufti with whom to exchange experiences . . . sing-songs . . . exultant chance relatives . . . curious passers-by . . . tiny tots tugging at restraining arms . . . laughter, shouts, cheers . . . even tears, as a Czech spies his national flag . . . all the pandemonium associated with a station reception.[21]

When the first seventy-five RAF airmen arrived at No. 41 SFTS, Weyburn, a civic greeting awaited them and "they were taken to the local Legion Club for coffee, sandwiches and speeches." More than a hundred more RAF airmen arrived the following week. On Christmas Day they ate a traditional turkey dinner at the base, after which they were "met by their prospective hosts and by the Boy Scouts and escorted to various homes in the city and entertained for the remainder of Christmas Day."[22]

Western hospitality could be embarrassing. Townsfolk often took RAF students for veterans, heroes of Dunkirk and the Battle of Britain – and in general the students, revelling in the attention, were in no hurry to set the record straight. "No wonder we had a rather high opinion of ourselves," recalls former FAA pilot Bill Martin, who took elementary flying training at No. 12 EFTS, Goderich, Ontario. "We persuaded those who had cameras to photograph us gazing at distant horizons in noble, narcissistic poses, or climbing out of aircraft cockpits laden down with Mae West life jackets and parachutes, as if we had just returned unscathed from a successful sortie. . . . We talked to the locals and the gorgeous, nut-brown college girls visiting from farm camps in the area as if we had personally held the Germans at bay in 1940 and were now on a sort of refresher course, so to speak, preparing for our return to finish off the Japanese. We were disgraceful phonies and had no shame."[23]

Occasionally, however, encounters with civilians were less gratifying for the students' egos. When RAF student Len Dunn went to New York City to celebrate his graduation from SFTS, a lady mistook his unfamiliar uniform for that of a hotel porter and demanded to be shown the way to the toilet. (Joe Hartshorn, an American instructor in the RCAF, went to Philadelphia to get married. Entering a flower shop, he was confronted by an angry owner, who wanted to know why he was late delivering the last order of flowers. Hartshorn had to explain that his blue uniform was not that of the local flower supply firm.)

Civilians living near training bases took a personal interest in *their* students, whatever the students' nationality. The presentation of pilots' wings to the first graduating class at No. 11 SFTS, Yorkton, in the summer of 1941, was front-page news, with every pilot named. Indeed the practice continued throughout the war, although by 1943 the lists of graduates' names no longer made the *Enterprise*'s front page. Townsfolk felt a sense of personal loss when students, instructors, or ground staff were killed or injured in accidents. Some funerals attracted astonishingly large crowds of civilians, very few of whom knew the casualties personally. The bodies of Canadians or Americans were usually sent back to their home towns; others were buried locally.

Understandably, relations between civilians and air force personnel tended to be closer when RCAF rather than RAF bases were involved. Canadian airmen could take an active interest in local baseball and hockey activities; few townspeople could drum up the same enthusiasm for the cricket and soccer beloved of the British and Commonwealth airmen.

The record of relations between BCATP airmen and civilians, though good, was not unblemished. Ken Fulton of Truro, Nova Scotia, remembers arriving at No. 3 ITS, Victoriaville, Quebec, and being warned that the local residents were not in favour of the war and tended to be unfriendly to airmen. Fulton discovered the truth of this a few days later when he was one of a group of airmen sent on a route march through the town. Locals spat at them. Bert Lee was a student observer at No. 8 AOS, Ancienne Lorette, Quebec, in the fall of 1942. He recalls: "Our orders were not to go into Quebec City, the nearest large town, alone or while in uniform. It was not safe." Ray McFadden, a ground staff wireless instructor, remembers "open hostility between the townspeople and the RCAF" at Valleyfield, Quebec. Jim McPhee of Oplir, Ontario, later an air gunner with 408 Squadron, also found himself at Valleyfield during his training. He says that "several incidents occurred between our airmen and civilians. We always went into town in groups." On the long trip westward to No. 33 SFTS, Carberry, Manitoba, RAF student Jeff Rounce was taken aback by the hostility of the locals during a stopover at Rivière du Loup, Quebec. He says, "When we discussed this with the train attendants we were told the French Canadians, who at the best of times were not over-friendly with English-speaking Canadians, were particularly bitter because they thought we had let the French down in France during the German invasion."

Quebec quickly developed a poor reputation, and most English-speaking airmen dreaded postings there. But when RAF student Doug Harrington was posted to No. 8 AOS, Ancienne Lorette, a pleasant surprise awaited him. He had travelled with three other LACs. When they got to Quebec City, they found the road to Ancienne Lorette completely blocked by a heavy snowstorm. The airmen struggled through the snow to the Air Force Club in the Place Montcalm. "As we arrived on the scene," recalls Harrington, "a local resident called in to say he had booked two rooms at the prestigious Château Frontenac for the use of stranded cadets . . . and four lowly LACs thus spent two nights there, hoping against hope that the snow ploughs would be unable to re-open the road."

Curiously, it was not in Quebec that the worst troubles between BCATP airmen and civilians occurred. Altercations between airmen and civilians were of course inevitable anywhere. Bob McBey remembers some hostility during his training at No. 39 SFTS, Swift Current, Saskatchewan, where local youths put airmen's backs up with references to "Limey bastards." A few black eyes would sometimes result, but, says McBey, there was "no harm done and some steam was let off." The worst trouble occurred in the unlikeliest of places: Moose Jaw, Saskatchewan, the home of an RAF base, No. 32 SFTS. Things came to a head in the late summer of 1944, but the ingredients of the mess had been brewing for some time. Ironically, in view of what took place there, Moose Jaw prided itself on being "the most British city in Western Canada" at that time and regarded the RAF airmen at the base as potential postwar immigrants. The city, with a population of over twenty thousand in 1941, had been less affected economically by the establishment of the air force base than many smaller centres in the province. The cause of the "troubles" was mainly at the school, not in the city.

It all began in the summer of 1943 with the arrival of a new commanding officer, Group Captain E. J. George. The new CO was far from impressed with what he found at No. 32 SFTS. His predecessor had been excessively permissive, he believed, allowing the airmen too many privileges. Without delay he withdrew sleeping-out passes for most single airmen and forbade the wearing of civilian clothes on pass. The measures could hardly be considered draconian in the context of a world war, but nevertheless the airmen reacted at once, holding strike meetings. The following morning about two hundred of them refused to go on parade. Instead, they marched around the camp, trying to encourage others to join them. They ignored orders to disperse and warnings that their actions were mutinous. Peace was at last restored when, rather surprisingly, the CO agreed to suspend the new regulations pending a review of the matter by RAF HQ. If the troublemakers thought they had won a victory, they were soon disappointed. Fifteen of the ringleaders were charged with causing or joining in a mutiny, and thirteen were found guilty at subsequent courts martial.

Sloppy discipline had become a way of life at No. 32 SFTS. It manifested itself in a number of ways, including aircraft serviceability, which was consistently 10 to 15 per cent below that of comparable bases. George's predecessor had lived in Moose Jaw with his wife and had devoted a great deal of his time to the development of good relations with the city and its citizens. In view of the problems the new CO inherited, it was hardly surprising that he elected to live on the base. As a result he had almost no contact with the city and its inhabitants. At the time he took over, however, there appeared to be little bad feeling between the townspeople and the air force, although some citizens were of the opinion that the standard of RAF airmen had declined since the base had opened in December 1940. The local young ladies weren't complaining, though. Which was precisely the problem: as far as the young men of the city were concerned, the air force men had become far too popular.

In September 1944 the situation in the city became tense, with the RAF personnel – apparently all ground staff – on one side, the young male citizens of Moose Jaw on the other, and the young women of the city in the middle. The chief constable put his finger on the problem. In a report to the RCMP, Regina, he stated emphatically (if not quite grammatically) that the "feminine gender appear to favor men in uniform preferably to those in civilian dress."[24] Air force investigators did their best to dismiss the young civilians as disreputable types "of the street corner, loafer variety mostly . . . affect[ing] a unique manner of dress which causes them to be referred frequently to as 'zoot-suiters.'"[25] A newspaper report described them as "irresponsible youths between 16 and 18."[26] Some RAF investigators tried to make the case that the troublemakers were outsiders sent in to foment riots.

It wasn't so. The RAF men probably created the ill feeling by behaving badly at dances and restaurants, elbowing their rivals aside, showing off their popularity with the women of the city, and generally making themselves unpopular. Few of the airmen liked Moose Jaw. They found the prairies a bleak wilderness, a cultural wasteland. Unhappily, they didn't have the courtesy and sense to keep their opinions to themselves. Neither did the No. 32 SFTS magazine, *Prairie Flyer*. Supercilious remarks about the area had become a familiar theme in the publication, with regular comments about the prairies being "dull and uninteresting," and "especially drab in winter." Indeed, what happened at Moose Jaw might well be an example of the unlikeliest of events: a BCATP newsletter contributing to civic upheaval. Admittedly, the *Flyer* did occasionally attempt to present a more positive picture of life on the prairies. One article recalled the welcome the RAF received from the citizens: "The people of Moose Jaw were, from the start, almost embarrassingly hospitable. I remember on Christmas Day several kind people called on the chance of picking up some airmen to entertain, only to find that there were none left to take." But the same writer couldn't resist remembering being "rather shaken to find that a place called Moose Jaw really was a town and not just a row of wigwams." [27]

The city's young men didn't help the situation when they called the RAF airmen "yellow-bellied English bastards" – presumably because they were not serving in a combat area. And they bitterly resented the servicemen's habit of asking the civilians why they were not in uniform.

On September 9, fighting broke out. The two sides armed themselves with a variety of weapons, including lengths of lead pipe, jackknives, aircraft cables, and homemade billies of assorted sizes. For the first few evenings, the incidents were little more than angry clashes between individuals or small clusters of men. Then, on September 12, more than three hundred airmen came to Moose Jaw, most of them planning to spend the evening at a dance at the Temple Gardens. While the dance was on, several groups of young men, teenagers mostly, attacked a party of eight airmen in Central Park, inflicting a few injuries, though nothing serious. When the news got back to the Temple Gardens, some two hundred airmen immediately left the dance and streamed into the park, looking for the assailants. Soon the search spread into the streets. Fights broke out in several spots, like flash fires igniting in a tinder-dry forest. Crowds quickly gathered, the *Regina Leader-Post* reporting that "thousand of citizens flocked to the downtown area hoping to get 'ringside seats'"[28] The police worked speedily and efficiently, containing the trouble principally by keeping the opposing groups apart. They arrested some civilians, but no air force personnel.

Peace descended on the city of Moose Jaw when all service personnel were ordered to return to camp. One of the minor mysteries of the affair is why Group Captain George didn't, at this first outbreak of real hostility, confine his men to the base. Whatever the reason, he took no such action. And the following evening the RAF returned to the city in force, "looking for trouble," in the words of Deputy Assistant Provost Marshal, No. 4 Training Command. The local police claimed that they saw officers and service police among the troublemakers, deliberately jostling and harassing the civilian population, but this has never been confirmed. Feelings ran high. The deputy mayor did his best, broadcasting over the local radio, pleading with the citizens to remember that the airmen were their guests. His words did little to ease the situation. Townsfolk were understandably angered. They had opened their homes to the RAF and this was how they were being rewarded. At last George imposed a curfew at the base. Police reinforcements arrived. Peace was once again restored. And afterwards? It can't be coincidence that No. 32 was the first SFTS in the BCATP to close, barely a month after the disturbances.

Perhaps the oddest element of the affair is that the RAF was involved, not the RCAF. As they demonstrated again and again, both at home and overseas, Canadians tended to be considerably less amenable to military discipline than the British. The first group of BCATP-trained wireless operator/air gunners, all Canadians, arrived in Britain early in 1941 and immediately ran into trouble with the RAF. Posted to Cranwell to bring their training received in Canada up to RAF standards, they found that no one seemed to know what to do with them. Soon the Canadians, every one of whom was a sergeant wearing an aircrew badge, were spending their days trudging around Lincolnshire on a series of totally pointless route marches. They objected so vociferously – and violently – that Canadian High Commissioner Vincent Massey became involved. He smoothed troubled waters by arranging for Canadian officers to take over. None of the "mutineers" was punished. Historians have repeatedly referred to Canadian servicemen as brash, quarrelsome types, quick to criticize orders or superiors – they considered stupid. Growing up in a relatively classless society, the Canadian serviceman was seldom as intimidated by rank as was his British counterpart. Yet it was British airmen who mutinied and later rioted at Moose Jaw, the only such incident in the history of the BCATP.

Students from other dominions had been arriving in Canada since the fall of 1940. Australians came first, aboard the liner *Awatea*. Distinctive in their dark-blue uniforms, they entrained in Vancouver for No. 2 SFTS, Ottawa, where they received a warm welcome from Canadian airmen who wanted "to show our Australian cousins how welcome they are in our midst."[29] The students from down under had already completed their elementary flying training; now they had advanced to the next stage of their training, learning to fly the spirited Harvard. Other Commonwealth airmen soon followed. Most seemed to like Canada and Canadians. New Zealander Ron Mayhill recalls:

I retain a kaleidoscope of vivid, happy memories: men wearing fur coats at the stations, riding up and down the first escalators we had ever seen, in a department store in Vancouver, the glorious Rockies with firs and telegraph posts bending under the snow, the long train with engines at both ends trying to catch its own tail up the inclines, and the friendly reception at Brandon. Brandon had military discipline, yet the instructors became friends. There are memories of ice-skating and being helped up by laughing girls, the staggering vastness of the prairies, the lakes a hundred miles north of Winnipeg, at Paulson, sixty degrees below freezing, frozen ears, the shock of double-glazed, heated rooms and venturing outside into the Arctic, the delight of many of our fellow New Zealanders and Aussies who had never seen snow before; and above all, the great comradeship which produced lifelong friendships.

In the main they were a cheerful, extroverted group. When Torontonian Bill Hutchins was a student at No. 8 SFTS, Moncton, New Brunswick, he found that the Canadians and Americans in his class seemed to stick together, with the Australians and New Zealanders forming another group. There was no animosity, however, just a lot of good-natured bantering, the North Americans taking particular delight in trying to copy the Aussies' accent – "To-die is pie die" (translation: "Today is pay day") being a favourite expression. However, the Australians also had trouble with the Canadians' verbal idiosyncrasies. Don Charlwood, an RAAF (Royal Australian Air Force) student observer, was puzzled when a Canadian NCO shouted good-humouredly to a squad of recent arrivals: "'Fall in, eh!' The addition of 'eh' made his order sound like a friendly suggestion. When we ignored it, he roared, 'All right, you goddam Osstrylians, *fall in!*'" [30]

The students from down under got along well with the civilian population. Don Charlwood rated the people of Edmonton "the most hospitable on earth." Ron Anstey of Wellington says, "We New Zealanders were welcomed into Canadian homes many times." Fellow New Zealander Alan Gibson remembers being in Toronto late in 1942: "We were having a beer and a meal when we met a man who owned a button factory, and he invited us to stay with him and his family for Christmas. I still remember his name – Ed Squire. . . . Without doubt one of the nicest Christmases I've ever had the good fortune to enjoy." Ray Lackey, a student pilot from Sydney, was low flying with his instructor near Calgary when he noticed a girl standing in the snow near a farmhouse. She held a sign reading: PHONE 307 STRATHMORE. Back at base, Lackey called the number and discovered that the "pulchritudinous Canadian lass," as the No. 3 SFTS base newsletter later described her, was thirteen-year-old Bernice Christiansen. "I went to dinner with Bernice and her family," says Lackey, "and from that day until I left for Europe, their home was my home." (Forty years later, Lackey met Bernice again when she and her husband visited Australia.)

Bill Harker of Lethbridge, Alberta, remembers the high spirits of the Australians and New Zealanders at No. 15 SFTS, Claresholm, Alberta, "who seemed to get great joy cutting down the flag pole." Harker adds that the problem was only solved when the pole was "encased in iron from the ground six feet up." The Aussies and Newzies often displayed a disdain for authority that surpassed the Canadians'. John Turnbull was at No. 3 Manning Depot, Edmonton, in 1941, when "a contingent of RAAF arrived one day and, denied passes after having been confined to ship and train for weeks, grasped the fire axes as they marched straight through their assigned huts and out the back doors. . . . They returned in a few days!" Ex-wireless operator/air gunner Neil Fletcher remembers an entire unit of Australians at No. 2 Wireless School, Calgary, refusing to march off the square to classes because of some issue. "If one was in trouble, they were all in trouble," he adds.

In the spring of 1943, a few Australians training at No. 7 B&GS, Paulson, Manitoba, took it into their heads to chop down the flagpole outside the station headquarters. Furious, the CO ordered everyone on parade. Ron Mayhill was there and recalls that armed guards flanked the serried ranks of students. Determined to put a stop to the practice of flagpole-pruning, the CO delivered himself of a fusillade of threats about the dire consequences of mutiny in wartime. He then demanded that the culprits own up immediately. Without hesitation, the entire contingent of more than a hundred Australians and New Zealanders stepped forward.

Supplied with shorts for summer wear, the Australians and New Zealanders tended to put them on when *they* felt they should be worn, no matter what anyone in authority had to say on the matter. In some rural areas, the shorts caused a stir. Wayne Nicholls from Auckland, New Zealand, trained at No. 2 AOS, Edmonton. One warm Sunday in May, he and a friend cycled out into the country for a picnic with two girls. On the way home, a police car stopped and pulled the New Zealanders over. The officer told the airmen to hurry back to the base and change, otherwise they would be arrested for indecency.

While Allan Turton was at No. 5 Manning Depot, Lachine, Quebec, early in 1942, a contingent of Australian trainees arrived in Montreal by ship. "A newspaper reporter saw one of them sitting on a park bench, looking rather glum. . . . The Aussie told the reporter that their kangaroo mascot had escaped when they were getting off the ship. The reporter published the story and contacted the CO at Lachine. . . . The Australians were given time off to hunt for the kangaroo," says Turton.

Bert Lee was at Lachine later that same year and was fascinated to watch the Australians catching and examining the first snowflakes of the season. New Zealander Keith Prior says he had never seen snow until he arrived in Canada. The brutal Canadian winter temperatures shocked the newcomers. Wayne Nicholls got his ears frostbitten while waiting for a streetcar in Edmonton. "It was fifty-two degrees below," he recalls, "but the cold was so dry I didn't notice it and did not have my earflaps down." Keith Prior, like many airmen from down under, found the BCATP barracks grossly overheated. The New Zealanders took to opening their windows at night, a practice which amazed and distressed the Canadians. First, why would the Newzies not want their barracks cosily warm at night? Second, didn't they realize that they were running the risk of frostbite? And third, didn't they know that fuel cost money?

While the BCATP operated primarily with Canadian, British, Australian, and New Zealand students, other nationalities made their presence felt in Canada. The Norwegian government-in-exile in London had had a number of American aircraft on order when the Germans invaded. Now, wanting to make use of them, they asked Ottawa for permission to set up a Royal Norwegian Air Force training centre. They got the Toronto Island airport, which soon became known as "Little Norway." Later the base was moved to Muskoka Airport. The Norwegians paid for their supplies and equipment and were never part of the BCATP, although they received a considerable amount of support in various forms. Most graduates of "Little Norway" received advanced training at BCATP schools before going overseas. The Norwegians had a strong impact on Toronto and Muskoka, flying blue Cornells, Hawks, and Northrop A-33s, and charming the local female population with their blond good looks and intriguingly exotic accents. Their popularity might have incensed some RCAF airmen, for in Toronto more than a few posters exhorting young men to join the RCAF had the sub-heading "AND WATCH THE NORWEGIANS FLY" scribbled by angry hands.

Another government in exile, that of French general Charles de Gaulle, discussed the setting up of a training establishment in Canada along the lines of "Little Norway." Ottawa had to turn the request down, amid some embarrassment, for the BCATP had no facilities for training aircrew in French. It would be 1942 before Free French airmen were trained in Canada, coming as part of the RAF quota, along with Belgians, Poles, Czechoslovakians, and others. Two prairie schools, No. 32 SFTS, Moose Jaw, and No. 34 SFTS, Medicine Hat, Alberta, became the multinational centres in Canada. But English remained the language of every course, no matter what the nationalities of the students.

This also caused problems for many Canadian-born students. One was Roger Coulombe, a young man from rural Quebec, who spoke little English when he enlisted in July 1941. He came from the village of Berthier-en-bas (now known as Berthier-sur-Mer). "Nobody in the entire village could speak English," he says. "All along the south shore of the St. Lawrence River, it was French only." He recalls having to wait three weeks for his uniform because his size was in short supply. "When at last I was called to the quartermaster's store to get mine, the sergeant in charge asked me, 'When did you enlist?' As I could not understand a word of what he was saying, he repeated his question differently: 'When did you swear?' I didn't understand those words, either, so he called a French-speaking airman who repeated the question in French. Then of course everything was clear to me and I was able to respond, '*Le 21 juillet, monsieur.*'"

At No. 3 ITS, Victoriaville, Quebec, Coulombe was with about a dozen French-Canadians, none of whom could speak or understand English. It took the authorities more than a month to realize the obvious: these airmen needed special assistance to prepare them for the highly technical training they were about to undergo. "We were posted back to No. 4 Manning Depot in Quebec City," Coulombe says. From Quebec the airmen journeyed to Fingal, Ontario, to learn English. Most of their three weeks there, they did nothing but send Morse messages. Then it was back to Victoriaville for ITS. Here, Coulombe remembers, there "were only a few French-speaking airmen, perhaps a dozen among a few hundred

English-speaking individuals. Sometimes, when we were talking French among ourselves, some guy from English Canada would shout: 'Speak white, you French frogs!'" (On the other hand, Lucide Rioux, an aero-engine mechanic who trained in Canada before going overseas, had no similar experience. He says, "I was supposedly French-Canadian, but I cannot recall being slighted in any way by any other 'Canadian.'") In those days, says Coulombe, "there was only one language in the air force, and it was English. Nothing was ever done in French. No instruction, no demonstration, nothing! You had to understand English, or else." Motivated by an overwhelming ambition to become a pilot, Coulombe worked tirelessly, never going out in the evenings but staying in to pore over the information imparted during the day. The other students helped him perfect his English.

Interestingly, he remembers no resentment among his fellow francophone students about the air force's language policy. "We took it for granted that the air force of the time was an English-speaking service . . . in fact, we felt guilty for not knowing more English." The going was tough for most francophone students, and inevitably some got discouraged and gave up, being remustered to other trades. Determined to become a pilot, Coulombe refused even to consider the possibility of failure. "I was not going to give up, no matter what!"

After ITS, Coulombe travelled to No. 11 EFTS, Cap de la Madeleine, Quebec, where he learned to fly Fleet Finch biplanes. During his training, Coulombe encountered a volatile French-Canadian instructor who persisted in screaming at him in incomprehensible English. "I wish he had been screaming in French. At least I would have understood him!" The instructor refused to talk French, apparently regarding it as a "forbidden language." Coulombe adds, "What a difference when I got to England and started training with British flying instructors. . . . They were good, they had class . . . and they never screamed any of their instructions." It was in England, assisted by many young ladies encountered in local pubs, that Coulombe found himself finally thinking in English, rather than having to translate everything. He says he always felt he was in friendly territory in England, whereas in Canada he sometimes felt that he was in a hostile environment. (Coulombe flew a tour of operations with 426 Squadron, earning the DFC – and an award for dropping more bombs on Berlin than any other pilot in Bomber Command.)

Another worry for the air staff was the supply of Canadian aircrew recruits. Demand was beginning to outstrip supply. The need was for young men in excellent health between eighteen and twenty-eight years old with two or more years of high school education. The country had approximately 105,000 young men so qualified. Some 15,000 to 20,000 more became eligible every year. But the army and the navy – and industry – also sought recruits from this group. It seemed clear that unless the air force's enrollment requirements were eased a manpower crisis was inevitable.

In April 1940, when BCATP training began in earnest, pilot trainees had to be eighteen but not yet twenty-eight years of age. The age limits for other aircrew candidates was eighteen to thirty-two. Five months later, the maximum age for pilots was raised to thirty-one. In January 1941, it was raised again, to thirty-three. The following October, the minimum aircrew age came down to seventeen and a half, with the maximum age for pilots now at thirty-three and up to thirty-five for other aircrew, with air gunners being accepted up to thirty-nine. At the same time, medical requirements were relaxed in the categories of blood pressure, vision, and heart.

Educational levels, too, were under review. At this period, recruits had to have achieved a junior matriculation level in order to get into aircrew. But the authorities could see that the requirement was costing them too many potential aircrew candidates. They had little choice but to begin considering "suitable" candidates who had not actually obtained their junior matriculation. It was an indication of how rapidly the exigencies of war had influenced the recruiting process. Little more than a year before, the air force had been able to choose among scores of applicants for every vacancy; only the most superbly qualified need apply. Now the situation had changed, and the air force had to compete for the best and brightest of the nation's young men.

Paradoxically, as the supply of recruits began to cause concern in Canada, trained aircrew were arriving in Britain in unprecedented numbers. Soon the personnel pipeline would be full to overflowing. Allan Turton landed in March 1943, having graduated as a navigator from No. 8 AOS, Ancienne Lorette, Quebec. Like many thousands of his compatriots before him, he found himself at Bournemouth. There he waited for a posting to a squadron . . . and waited . . . and waited. He remained at the south coast resort town for almost five months. Eventually he received orders to journey to Dumphries, Scotland, for a refresher course in navigation, which by then he badly needed. From there he went to Honeybourne OTU – "very pleasant in the beautiful Cotswolds" – where he crewed up with a Canadian pilot and bomb aimer and an RAF wireless operator and rear gunner. Like so many OTUs, Honeybourne operated ancient aircraft of dubious serviceability – in this case Whitleys, outdated twin-engined bombers. "Two blew up in the air," Turton recalls, "because of a generator short circuit that ignited petrol vapour in the starboard wing. The rear turret of one aircraft landed in the street in the village of Broadway." Shortly after Christmas, Turton's crew was posted to 432 Squadron, RCAF, but it was not until February 25, 1944, that they set off on their first operational sortie. The target was the M.A.N. diesel works at Augsburg. "After dodging some flak . . . we made a nice bombing run in perfect visibility and set course for home." But the crew made a mistake typical of "sprogs," according to Turton, flying "straight and level at 22,000 feet, in an avenue of fighter flares." A night fighter slipped beneath the Halifax and "fired his cannons up through our empty belly and blew the middle out of the aircraft." Turton and four other members of the crew escaped by parachute and spent the rest of the war in POW camp.

It is a reflection of the huge numbers of aircrew now flowing through the BCATP that Turton had completed his navigation training at No. 8 AOS in January 1943, yet it was not until February 1944, more than a year later, that he went on his first (and only) op.

CHAPTER SEVEN

Over Here

Airspeed Oxford

"This generation of Americans has a rendezvous with destiny"
– Franklin Delano Roosevelt

While Europe went to war, America still grappled with its economic problems. A full decade after the onset of the Great Depression, the country's economy remained stuck firmly in second gear. In 1937 the market had slumped, sending shock waves along Wall Street. Was another worldwide collapse in the offing? Was the system going to break down again with the same grim consequences? Was this the right time to sell? Or buy? Or wait and see? Pundits monitored the economy's progress like doctors at the bedside of a frail patient. More than ten million Americans looked for work.

As far as the war in Europe was concerned, many Americans saw Hitler's ambitions as a direct threat to the United States, and they favoured firm action. But many more wanted no part of it, remembering bitterly the experience of World War I. The unspeakable slaughter seemed to have served no more worthy cause than bigger profits for the industrialists. If the Europeans wanted to fight each other again, let them get on with it. Even the Japanese invasion of China failed to make much impact on American public opinion, at least initially.

During President Franklin Delano Roosevelt's first term in office, from 1933 to 1937, the United States had taken a strictly passive role in international affairs. When Italy invaded Ethiopia in 1935, Congress's response was to pass a neutrality act demanding that the president take no action that might upset either of the belligerents. The outbreak of the civil war in Spain in 1936 precipitated only a firm restatement of America's neutrality and an order forbidding the shipment of arms to either the Republicans or the Nationalists in that tormented land. During Roosevelt's second term, 1937 to 1941, Americans became even more isolationist. Congress passed another neutrality act in 1937 and forbade U.S. citizens to travel on vessels belonging to belligerent nations or to sell supplies to those nations.

It was a difficult time for Roosevelt. Although convinced that a major war was coming and that the United States would eventually become involved, he was also keenly aware of the political dangers of saying so. He could do little more than keep warning Americans of the crisis facing them. In October 1937 he informed an audience in Chicago that international law and order were in danger from a disease of the mind and soul. He urged the institution of a "political quarantine" of those threatening world peace, adding that non-involvement was no way to guarantee peace. Somewhat illogically, the isolationists saw this as further proof of his "war-mongering"; political rivals accused him of making use of the international crisis to divert attention from the failure of his domestic policies, in particular his inability to bring a semi-moribund economy back to life. Didn't the nation have enough of its own problems without taking on everyone else's? Americans condemned Hitler's persecution of the Jews, but only a small minority wanted the United States to take any action. Even the idea of imposing an economic embargo on Germany met with no enthusiasm. When Germany invaded Poland, Roosevelt used one of his radio "fireside chats" to tell his listeners, rather ominously, that he would maintain American neutrality as long as possible but that the situation called for the strongest possible condemnation of what he referred to as "international banditry." If he expected the population of the United States to rise up as one and declare its loathing of the Nazis' actions, he was disappointed. Opinion simply polarized. Isolationists saw the events in Europe as further proof of the need to stay out, while "internationalists" pleaded anew for strong action to halt the aggression.

It was the time of the America First Committee and Charles Lindbergh, the boyish, smiling hero of the twenties, now balding and serious, shaking his fist at gigantic crowds as he warned them of the dangers of getting involved in Europe's mess. Britain couldn't win, he told his listeners, and arming her would do nothing but exacerbate an already delicate international situation. Lindbergh, seen by many as the man to oppose Roosevelt in the 1940 election, had visited Nazi Germany in 1936 and was just as impressed by what he saw as Mackenzie King was a year later. Both men responded to the well-ordered society, the obedient, productive masses, the absence of an intrusive, abusive press corps (particularly appealing to Lindbergh, who had suffered so much at the hands of the media). Lindbergh fell under the spell of the Nazis and believed them capable of beating any foe.

But although the isolationists and internationalists made a lot of noise, most Americans belonged to neither camp. They simply hoped for the best. The apparent stalemate in the early days of the war – the "Phoney War" – reinforced the feeling that this was no time to be taking sides. Things seemed to have settled down after Poland. With a bit of luck the politicians would be able to patch things up; there seemed to be a better-than-even chance that the war would be settled diplomatically. Most Americans felt that the war had been started by crooks on both sides and that soldiers were being asked to die for slogans as hollow as those of 1918.

Roosevelt's political foes charged that he talked about peace while he was busy arming Britain. His aim was vast, unprecedented personal power, they said; unless checked, he would attempt to become the first American dictator.

Despite such sentiments, as the 1940 election drew near it became apparent that the Republicans had much the same view of the Nazis as did Roosevelt's Democrats. The Republican candidate, Wendell Willkie, was a staunch supporter of aid to the Allies and of strengthening U.S. defences. After the stunning defeat of France, even the most die-hard isolationist had to agree that the situation had become far more perilous for America. If Britain went the same way as France (a certainty, according to the U.S. ambassador to Britain, Joseph Kennedy) the United States would stand alone. Roosevelt lost no time in urging changes in the U.S. neutrality legislation, recommending major additions to the country's defences and increasing aid to Britain and other friendly nations. Congress agreed. Massive outlays were appropriated for updating and enlarging the U.S. armed forces. The recently passed neutrality acts were amended to permit arms shipments to Britain; thus did the United States become the "great arsenal of democracy." [1]

Interestingly, signs of a growing realization of the awful inevitability of war could be read into the 1940 and 1941 advertisements of major corporations who were "helping America prepare." None stated precisely what they were preparing for, but only the most obtuse of readers could have had much doubt. It was during this uneasy period that Billy Bishop visited Washington and called on the president. Bishop's son, Arthur, remembers his father telling him that the president said it would be a good idea to paint black crosses on a Harvard and send it up to Boston to drop a few bombs to awake the citizens to the dangers around them.

Despite spending much of the fall of 1940 in Washington and doing remarkably little campaigning, Roosevelt won the 1940 election easily. It was a shattering blow to the isolationists. Although they maintained their attacks on the administration, the election had made one fact crystal clear: most Americans now sided with the president – that "soothing simplifier, manipulator of truth, drawer of false analogies," [2] as the isolationists referred to him. The name-calling continued. To the isolationists, Roosevelt's supporters were warmongers and agents of imperialist Britain. To those who sided with Roosevelt, the isolationists were fascists.

Then came Lend-Lease and the draft. The former gave Roosevelt power to spend billions on arms and other supplies, and sell, transfer, lend, or lease them to nations whose defence was considered vital to the defence of the United States. Britain and China were the first beneficiaries of the plan. Not surprisingly, the isolationists fought the bill with all their might. "The very title of the bill is a fraud," huffed Senator Robert A. Taft. "Lending war materiel is much like lending chewing gum. We certainly do not want the same gum back." [3] A Congressman of Irish descent came up with new words to "God Bless America": "God save America, from British rule" [4] – although it's hard to think of anything more unlikely than a British attempt to take over the United States during World War II. Another legislator suggested that the bill include the words: "Nothing in this act shall be construed to authorize or permit the President of the United States to lease, lend, or transfer the original Thirteen Colonies to King George of England." [5]

The first peacetime draftees in the history of the United States donned uniform in 1941. The army planned to induct nearly a million men in the first thirteen months. It proved difficult to equip them, however. Although American factories were busy producing all the paraphernalia of war, much of it was going overseas. It would be many months before the new generation of American soldiers had the means to fight. For now they did little more than wait. And complain.

Some didn't wait to be drafted; they headed north to volunteer for the Canadian forces, to the dismay of many of their compatriots. A few months after the outbreak of war in Europe, *Newsweek* wrote of a Los Angeles couple who complained to the FBI that their son had left for Canada "after being signed up by a 'foreign agent' for war service abroad." The magazine reported that "Canadian officials were said to have declared privately to newspaper men that 10,000 to 15,000 Americans were either in army ranks or on waiting lists, with 2,000 on the Royal Canadian Air Force list alone."[6] Although many Americans were violently opposed to their compatriots fighting for the Canadians, most seemed to approve of the existence of the BCATP, seeing it as part of the total defence system of the North American continent. One editorial described it as "democracy's first, long-term, planned offensive against totalitarian conquest." The story added: "The British Empire is staking its existence on this program; Americans should watch it carefully, for someday it may help us guard this hemisphere."[7] Most of the major magazines of the day carried stories about the vast training scheme north of the border, with scores of pictures of Harvards and Tiger Moths and pink-cheeked young airmen snuggled into fleecy flying gear. Sometimes the desire to raise eyebrows prompted the inclusion of data of dubious accuracy; for example, "the life span of a rear gunner in a modern bomber during combat is forty-seven minutes."[8] And occasionally over-enthusiasm puffed the prose: "You get the feeling . . . that the energies of the Dominion's 11,000,000 people have been turned, like a strong and silent wind, to flow in one direction. It is heartening to anyone who believes that Democracy can defend itself."[9]

Discussing the war, *Collier's* magazine expressed what was probably the view of most Americans. The journal said that it would "hate to see the United States get into this one," but took an unruffled view of volunteers heading north.

Young Americans are slipping into Canada in considerable numbers at this writing to join up for war service of
various kinds . . . and we can't rev ourselves up to shudder
about it the way some people can. . . . To us, the flow of
American volunteers across the border up Canada way is
mainly a sign that Americans by and large are still a
pretty good breed – which wouldn't seem to be anything
to worry about. So let's drop the shudder-shudder stuff
on this topic. Let's keep it as easy for the boys to scram
out and enlist as it is now; let's make it easier for them by
removing all doubts about recovery of United States citizenship when they come back, if they do come back. And
let's keep it as clearly understood as it is now that our
government is in no way responsible for anything that
may happen to them under other flags. [10]

In all, nearly nine thousand U.S. citizens served with the RCAF
during World War II. They were colourful, those volunteers –
professionals and playboys, convicted felons and husbands on the
run, idealists and mercenaries, kids seeking adventure, youngsters
seeking nothing but an opportunity to fly, middle-aged men
looking for work – and to all of them, the RCAF's need was their
golden opportunity. There was good-looking Harold "Whitey"
Dahl, an experienced pilot who had parted company with the U.S.
Army Air Corps some years before, following a brush with the law
involving gambling debts and bouncing cheques. Shot down during the Spanish Civil War, he found himself on trial for his life.
The court declared him guilty of rebelling against Franco and
condemned him to death. Only a personal plea to Franco from
Dahl's wife, Edith, saved the pilot. "Whitey" stayed in a Spanish
prison for a year after the Nationalist victory, and when he finally
returned to North America he lost little time in heading north to
Canada to join the BCATP as an instructor. Charlie Dow, scion of
a wealthy Jamestown, New York, family, arrived at the recruiting office in a brand new baby-blue Cadillac convertible; he

continued to use the vehicle when he was an LAC at Trenton. By contrast, George Harsh had served on a Georgia chain gang after narrowly escaping the electric chair following his conviction for murder. Harsh saved a fellow prisoner's life and was released, soon volunteering for the RCAF and becoming an air gunner. Shot down over enemy territory, he was sent to Stalag Luft III, North Camp, and became the security chief for the famous Great Escape because of his experiences as a convict. He is remembered by fellow airmen as "a true gentleman." A young American recruit named Bill Orendorf appeared on parade at manning depot with a medal ribbon on his newly issued uniform and incurred the wrath of the drill sergeant. "*That man!* What d'you think you're doing, wearing that ribbon?" It transpired that the ribbon was the *Croix de Guerre* and Bill had every right to wear it. He had driven for the American Ambulance Corps in France in 1940 and had been decorated for his heroic services. Another U.S. recruit was a remarkable young man named George Smith, who took an IQ test and achieved genius-level marks. The authorities wanted to commission him and put him into some sort of psychological work. He turned down their offer; he wanted to become an air gunner. He did – and died in the crash of a Halifax bomber. Maurice Phair from Limestone, Maine, became a pilot in the BCATP. Over Turin he ordered his crew to bale out of his damaged bomber, then found the aircraft still functioned and flew it back to England singlehanded. He was later lost during a raid on Essen in January 1943.

Some Americans volunteered simply because they wanted to fly. Burly Joe McCarthy grew up in New York, in what was then known as the Irish section of the Bronx. The family had a summer place on Long Island. Fascinated by all things aeronautical, McCarthy worked at odd jobs to pay for flying lessons at nearby Roosevelt Field, the airfield from which Lindbergh took off on his epic New York-to-Paris flight in 1927. McCarthy had on three occasions attempted to join the Army Air Corps; each time he had been told that he would hear from them. He never did. A neighbourhood friend, Don Curtin, had a job as a cruise director for the Holland America Steamship Company, but lost it soon after the war broke out in Europe. McCarthy bumped into him one day and the two began to explore the idea of travelling north and flying with the Canadians. "Within two days," says McCarthy, "we were boarding a bus and heading for Ottawa. We crossed the St. Lawrence via ferry, and the customs people helped us get the next bus to Ottawa.... We spent the night at the YMCA and the following morning proceeded to the air force recruiting office." Here McCarthy and Curtin ran into a situation that confronted many American volunteers: "We were advised to return in six weeks." McCarthy explained that they didn't have the funds to wait around in Ottawa; if the air force wanted them, it had better decide that day. The same evening, "we along with another fifteen recruits departed by train for Toronto and the manning depot."

Lucien Thomas of Richmond, Virginia, came to Canada at the beginning of the war. "My motive," he says frankly, "was adventure." He joined the army first, later transferring to the RCAF to become one of the members of the first "straight air gunners" course (originally all air gunners had radio training and were classified as wireless operator/air gunners). Of the twelve students in Thomas's course, only two survived the war. Thomas flew a tour of operations with 405 Squadron of Bomber Command and won the DFM. Another Virginian who travelled north was Clyde East, who had wanted to fly since he was a boy. He crossed the border at Niagara Falls and promptly hitched a ride with a navigation instructor from No. 33 ANS, Hamilton, Ontario, who not only took him to the nearest recruiting office, but also briefed him on his future career.

Some volunteers went to extraordinary lengths to get into the RCAF. Bruce Betcher of Crookston, Minnesota, was a nineteen-year-old freshman at the University of Minnesota who found his university studies acquiring a certain irrelevance compared to the great events taking place in the world. Sitting in a classroom just wasn't enough. He wanted to fly. But, partially colour-blind, he knew that he stood no chance of getting into the Army Air Corps. Over the objections of his parents – he was an only child – he decided to try the RCAF, reasoning that the Canadians might not be so fussy about minor medical problems. In May 1941, Betcher and a friend, Paul Hagen, climbed into a 1929 Packard and headed for the recruiting office in Winnipeg. They hadn't gone far when an unassailable truth hit them: they had no chance of making it. With the blithe optimism of youth, they had started off on their trip without checking three vital factors: the car's rate of consumption of fuel (some eleven miles per gallon), the cost of the fuel to get to Winnipeg, and their available cash. No matter how they calculated it, they couldn't reach their destination. Frustrated, they returned to Minnesota. Betcher vowed to try again. In July he did, this time in the company of another local friend, Leonard Isaacs. They reached their goal, but Betcher received disturbing news. Although the Canadian MO told him his colour blindness wasn't serious enough to prevent his service as aircrew, other problems had surfaced: a deviated septum and the absence of a layer of fat between the esophagus and the heart. Betcher immediately returned to Crookston and had an operation to cure the deviated septum; he also obtained a statement from a heart specialist declaring that his heart was normal.

On Friday, September 15, Betcher and Isaacs returned to the Lindsay Building in Winnipeg. The two volunteers passed their medicals with ease, then promptly ran into the same problem that had confronted McCarthy and Curtin in Ottawa. The air force wanted them to return at a later date. "We told the recruiters we had no funds, and if they wanted us they had to take us now or forget about it," Betcher recalls. Later that day Betcher and Isaacs were in the air force and on the train to Edmonton.

In common with other American volunteers, Betcher and Isaacs were not required to take an oath of allegiance to King George VI, since this would nullify their U.S. citizenship. They simply declared their willingness to obey their superior officers. At No. 3 Manning Depot, Edmonton, Betcher found service life as rough and ready as he had anticipated. The officer of the day strode through the Airmen's Mess asking, as tradition demanded, "Any complaints?"

Betcher piped up: "Cores in the apple pie, sir."

Aircraftman Betcher spent the next two days coring apples.

After a spell of guard duty, Betcher found himself at No. 4 ITS, Edmonton. Here, the medics examined the recruits even more thoroughly than at manning depot. To Betcher's dismay, they rejected him for aircrew; his colour blindness had again caught up with him. He was sent to Trenton, Ontario, the central base of the BCATP and the place where all aircrew "washouts" went for reassignment. It was an unhappy time for Betcher. In company with hundreds of other aircrew rejects, he witnessed a "hollow square" ceremony, in which an airman who had been found guilty of desertion was marched onto the parade ground to the sound of muffled drums. The adjutant cut away his cap badge, buttons, and belt buckle, after which he was marched away to his punishment. It was an effective ceremony. "I couldn't have been more impressed if they had hanged him on the spot," says Betcher half a century later. In the inexplicable way of military organizations the world over, the RCAF decided to test Betcher's eyesight yet again. Glumly anticipating the worst, Betcher prepared himself for the ordeal: "An elderly RAF flight lieutenant with World War I ribbons asked me to identify the colours on a Player's cigarette box. I did so and was cleared for flying duties with a note in my logbook which read, 'Colour blind safe.'"

Although he was officially cleared for flying training, Betcher had to use his wits to compensate for his partial colour blindness. On one occasion, he and Jack Simpson, his instructor at No. 12 EFTS, Goderich, Ontario, were circling the field before setting course on a cross-country exercise when flying control fired a flare. Betcher saw it but had no idea what colour it was.

"What does that mean?" Simpson wanted to know.

"What does what mean?" Betcher responded, pretending not to have seen the flare.

"They just fired off a flare."

"What colour was it?"

"Red," responded Simpson.

"That means immediate recall," said Betcher.

"Attaboy," was Simpson's comment.

Betcher travelled to Brantford, Ontario, for his service flying training. At No. 5 SFTS, he was a member of Course 59 and was soon introduced to twin-engined Ansons, which he found "steady, reliable, totally comfortable." His confidence in his flying ability grew. Indeed, like many students, he began to feel that he had learned most of what there was to learn about flying. "We felt we were hotshot pilots at this stage . . . described perfectly by that old saw, 'After forty hours a pilot thinks he knows all about flying. After four hundred he *knows* he knows all about flying. After four thousand he begins to realize he knows nothing about flying.'"

For the rest of his service career, Betcher encountered no more problems with his vision.

Some U.S. volunteers were too old for aircrew, but few permitted such trifling details to hinder them. Goronwy Pryse Lloyd of Schenectady, New York, was about ten years older than the average student in the BCATP but was determined to follow in the footsteps of a brother who had flown with the RFC during World War I. The younger Lloyd had to doctor his birth certificate to get past the recruiting officers. His youthful appearance helped him on his way. He was accepted. When he won his pilot's wings at No. 16 SFTS, Hagersville, Ontario, his brother came to the parade to pin them on.

In many cases volunteers' motives were admirably idealistic. Phillips Holmes, whose mother was Canadian, abandoned a successful acting career in Hollywood to join the RCAF. He followed in the footsteps of his brother Ralph, who was already a member of the air force in Canada. Ralph's wife was the popular singer

Libby Holman. The good-looking Phillips had appeared in a number of films, including *His Private Life* and *An American Tragedy.*

The plight of Europe's children motivated Hubert Clarence "Nick" Knilans to volunteer for service with the RCAF. He grew up in the farming community of Delevan, Wisconsin, in the grindingly hard times of the thirties. Photographs of refugee children in France and Belgium haunted him. He knew that something had to be done about the Nazis. In the fall of 1941, after helping the family with the harvest, he packed a bag and announced that he was off for a vacation in Chicago. While there, he decided to look into the possibilities of getting into the RCAF. "With my last five dollars, I bought a bus ticket to Detroit ($4.50), a sandwich (40 cents), and a bus ride to the border at Windsor (10 cents). The immigration officer showed me the way to the RCAF recruiting office. They took me in and sent me to a private home until my high school diploma and birth certificate could be sent for and approved. They gave me two dollars a day." When his documents arrived, the air force accepted him and he travelled to the manning depot at Toronto's CNE grounds. "Our contingent of thirty marched into the Sheep Shed sleeping quarters. Loud cheers greeted us," he remembers. It seemed to be an encouraging start to his new career. Then Knilans paid closer attention to the shouts. Suddenly he understood them. "They were shouting, 'Sucker! Sucker!'"

Joe Hartshorn of Chambersburg, Pennsylvania, was a student at Harvard while the Battle of Britain raged in the summer of 1940. Hartshorn had family connections with Britain and felt an intense desire to help defend the beleaguered island. Like so many young men of the period, he was as keen on flying as fighting Nazism. Thus, when a Canadian friend at Harvard told him that the RCAF was accepting American volunteers, Hartshorn lost little time in heading for Montreal and the nearest recruiting office – where an immaculately attired army NCO did his best to entice him to become a soldier, assuring him that he would get into action much sooner in khaki than in air force blue. Hartshorn declined the offer.

The technicalities of enlisting were seldom permitted to get in the way of a determined volunteer. Len Morgan went to the recruiting office in Windsor, Ontario, and found the enlistment process simple. "It wasn't slapdash, but it was quick. You had to show them your high school diploma and two letters of recommendation. If you showed up without them, the squadron leader would send you downstairs to see the sergeant and he would tell you about a little print shop down the street. They could make you up anything, from a letter of recommendation to a Harvard degree." Morgan recalls a young man from Cleveland who arrived without documents and availed himself of the print shop's services. When he returned to the recruiting office, the squadron leader said, "Oh, I see you found your papers." They were still wet, Morgan writes. [11]

Few American volunteers had reason to complain about their treatment by the RCAF. Some even believe that in many respects they were better treated than Canadians, particularly in the awarding of commissions at the conclusion of training. Certainly, their transgressions during training often seemed to be handled more generously than those of Canadian students. Bill Wagner, a volunteer from Akron, Ohio, recalls that in his class of forty-five at ITS, no less than thirty-eight were U.S. citizens – and the majority of those were Texans. The class became known as the Royal Texan Air Force, its members regularly wearing black tooled-leather cowboy boots with their uniforms. Bill Irvin of Lake Placid, New York, recalls the celebration that marked the end of elementary flying training at No. 13 EFTS, St. Eugene, Ontario, where he trained with Nick Knilans: "We lifted a Canadian pilot named Lucien Roy into a Fleet Finch, got him off the ground – blind drunk. He circled the airfield a couple of times before crashing into a hangar." Roy survived, but the Finch didn't.

On another occasion, Knilans, Bill Irvin, and a Canadian student named Andrew McCaughey were drinking beer in the canteen after closing time. The CO, a group captain, found them and sent them to the brig to cool off overnight. But no mention of the incident appeared on the airmen's records. Irvin later took advanced training at No. 2 SFTS, Ottawa, where he was required to do a long solo cross-country flight. "I made up my log beforehand and instead of following the flight plan, I flew to Lake Placid [his home town] and did aerobatics . . ." Irvin chose an unfortunate time to buzz the golf course, for King Peter of Yugoslavia happened to be enjoying a game there when the yellow Harvard came streaking over at treetop level.

Nick Knilans was also spotted indulging in illegal low flying. "On a solo flight west of Ottawa, I was beating up a sailboat on a secluded lake, trying to tip it over with my prop-wash. Another Harvard with two occupants appeared beside me. I realized that the one in the rear seat was an instructor." Faced with this unwelcome development, Knilans exhibited the sangfroid which would stand him in such good stead later in his operational career. He returned to base immediately and "asked a friendly mechanic to remove the engine hood and bang on the carburetor or anything else handy." He then hurried up to the control tower to report that he had experienced engine trouble and had been obliged to fly low over the lake. It is questionable whether anyone believed the story, but Knilans was not censured, despite the air force's strong feelings about low flying.

When the Japanese attacked Pearl Harbor, many of the American students in the BCATP wanted to transfer to the U.S. forces. It was a time of confusion and uncertainty. Rumours persisted that Americans in the RCAF who didn't volunteer to transfer to the U.S. forces might lose their citizenship. Section 401 (c) of the Nationality Act of October 14, 1940, (54 Stat. 1169) did in fact make unauthorized service in foreign armed forces grounds for loss of citizenship. But in April 1942, an agreement between Canada and the U.S. amended the act. Now persons who had served in the Allied armed services and had lost their citizenship as a consequence could regain it by taking an oath of allegiance before a U.S. diplomatic or consular officer. The previous month an agreement between Canada and the United States had been signed, facilitating the transfer of U.S. citizens from the Canadian to the U.S. forces. [12] The problem lay in getting all this information through the administrative strata of a largely disinterested foreign bureaucracy to the individuals concerned, particularly those serving overseas with the RAF. Joe McCarthy had just joined 97 Squadron at Woodhall Spa, Lincolnshire, and the message never did get through to him. On the other hand, Joe Hartshorn was completing his flying training in Canada and recalls that his superior officers told him he had to transfer if he was not to lose his citizenship. He eventually switched to the U.S. Air Force, but not before he had completed a tour with 419 Squadron of Bomber Command and had won a DFC.

Most U.S. units were glad to welcome the ex-RCAF fliers, particularly those with combat experience. But there were exceptions. Bill Irvin applied to transfer from the Canadian to the U.S. air force "largely for financial reasons." But he was surprised by the reception in the USAF. "Did they want us? Emphatically no! I recall very vividly the officer in charge telling us: 'You people in the RAF and RCAF have no sense of discipline. If you want to transfer, we have to take you, but we're not happy about it.'" After this heartwarming welcome, Irvin found himself on staff pilot duties for months, until he finally got into combat, in command of a B-26 medium bomber. When the transfers began, NCO pilots and navigators were given immediate commissions in the U.S. forces, but later on many were rejected, perhaps because of bad experiences with some of the original transferees. NCO air gunners were invariably welcomed, however, particularly those with combat experience.

When the BCATP had come into being, it was clear that the ambitious plan would require the services of far more professional pilots than Canada could provide. Could the RCAF hire American pilots? Would there be political problems? What about the individuals' citizenship? Billy Bishop remembered how thousands of Americans had come to Canada to join the RFC and RNAS during World War I. Wasn't it probable that the same thing would happen this time? Large numbers of American pilots were looking for adventure, or work, or both. What was needed, Bishop reasoned, was a system to interview and screen applicants in the United States, then spirit the successful candidates across the border. At the time, Bishop was on the reserve list, and he acted largely on his own initiative, although it seems probable that he had the unofficial approval of the minister of national defence. On September 4, 1939, the day after Chamberlain declared war on Germany – and when Canada was still at peace – Bishop had telephoned an old friend, Clayton L. Knight, an American veteran of the RFC. A well-known aviation artist, Knight had illustrated the famous book *War Birds: The Diary of an Unknown Aviator* and, in fact, was the "Clay" mentioned in that work. Knight had excellent connections in the American flying fraternity; at the time of Bishop's call he was attending the Cleveland air races, a house guest of Ohio's attorney general, Tommy Herbert, himself a World War I pilot. Bishop sounded uncharacteristically subdued, Knight remembered, as if weighed down by the thought of the job facing Britain and the Commonwealth. Knight recalled Bishop telling him that history was sure to repeat itself: "'American boys will want to help Canada as they did in the first war. But this time they must have direction and be screened. We need someone in the States to sort them out before they come across the border. You are in an ideal position to take on the job.' Then, apologetically he continued, 'We have no fund for this sort of thing, but that may be worked out someway – later.'" [13]

Knight's host, the attorney general, "had his fingers crossed a little on the idea" because of the legal and political implications of recruiting U.S. citizens for foreign wars. [14]

Knight decided to call the U.S. State Department to check on what was legal and what wasn't. He dialled the number, but when a voice came on the line it announced: "This is the German Embassy." "Brother!" Knight said afterwards. "That was the wrongest number in history." [15] When he began to make enquiries among the racing fliers, Knight found "a wide variety of opinions. . . . Clyde Pangborn, a 'round-the-world' pilot, who wasn't given to making snap judgments, thought that a lot of young fliers would probably be enthusiastic about going." [16] But Major C. C. Moseley, another veteran of World War I, now busy running a flying school at Grand Central Airport in Glendale, California, was dubious, believing that times had changed and only a few would want to go to Canada. On the whole, though, Knight was encouraged by initial reactions, particularly among younger airmen. Some contacts "gave him a kindly warning about the illegality of the enterprise," [17] but Knight pushed on.

Meanwhile, Bishop had been looking for someone to help Knight in his enquiries. He found him in another comrade, Homer Smith, a Canadian who had been a naval pilot in World War I and had inherited an oil fortune and lived in New York and Palm Beach for many years. Sworn into the RCAF with the rank of wing commander, Smith immediately headed south, joined forces with Clayton Knight, and together they flew from city to city, talking to likely prospects among the flying fraternity. Knight described Smith as an "outsized Canadian . . . an impressive, handsome, likable, very gregarious man of vision, with executive ability of an unusual sort. His connections in social and financial circles were wide and," Knight added, no doubt with the best of intentions, "he was more American than most Americans." [18]

The Clayton Knight Committee, as it became known, got off to a disappointing start. Bishop's first instructions were to "hold everything!" The venture was already in trouble. The Canadian

prime minister, Bishop explained, had become increasingly nervous about American political reaction to the plan.

Knight and Smith went ahead anyway, working in a kind of
limbo for several weeks, talking to interested pilots and outlining
the expected needs, the requirements, the limitations, but
promising nothing. In Ottawa, the ever-cautious Mackenzie King
agonized over the political ramifications. What about the American neutrality act? How would the American public react if it
became widely known that Canada was signing up U.S. pilots?
Might the Senate criticize Canada? Was the Opposition at home
likely to leap into the fray? Weren't things difficult enough without this added complication? Eventually the catastrophic turn of
events in France forced King to approve the recruiting of American pilots. In Britain, Churchill had taken over from the ailing and
increasingly ineffectual Chamberlain. An invasion was expected
daily, and Britain made it clear that the greatest assistance Canada
could render was to accelerate the entire air training plan. And
that could only be done with a flow of pilots from the United
States.

Still concerned, King asked the Canadian ambassador to the
United States, Loring C. Christie, about the probable reaction of
the American government. In response, Christie despatched a
message from the "highest quarter," assuring King that U.S
authorities would not be upset if their citizens went north to
volunteer.

Smith and Knight journeyed to Ottawa to attend a meeting of
the Air Council. "Present were the heads of all of Canada's commercial air lines, bush pilot groups, as well as top administrators
of the Commonwealth Air Training Plan, and United Kingdom
representatives. I was the only foreigner," Knight wrote. "The airlines were asked how many pilots they could spare for the BCATP;
each representative responded, but the total was a disappointing
seventeen." Knight spoke up: "Homer Smith and I have an initial
list of slightly over three hundred pilots who are willing to come

up to help." The reaction, according to Knight, was one of "pro-found shock."[19] Some delegates expressed doubt as to the fitness and suitability of many of the American pilots, but in the end they all agreed: Canada needed the pilots, so Smith and Knight were asked to recruit them.

In New York, Smith rented a suite in The Waldorf-Astoria hotel, declaring in his expansive way, "There's no sense in attempting to hide away in a hole-in-the-wall. It can't be kept a secret with the numbers of applicants we'll be seeing."[20] Before long he had opened branch offices across the country, in the best hotels in Spokane, San Francisco, Los Angeles, Dallas, Kansas City, Cleveland, Atlanta, Memphis, and San Antonio, each with a manager and a small staff. By this time the RCAF had opened a revolving bank account in Smith's name to cover the soaring expenses.

In Washington, Smith and Knight met General H. H. "Hap" Arnold, chief of the Army Air Corps, and Admiral J. H. Towers, chief of the Naval Bureau of Aeronautics, both old friends of Smith. The senior U.S. officers seemed relieved when they realized that there would be no effort to enlist men already serving in the U.S. forces. Arnold said, "According to the rules I'm working under, if a flying cadet gets fractious, goes in for low stunt flying, gets drunk one time, or we discover he's married, we must oust him. And if I was fighting a war, they're the kind I'd want to keep ... but I can't." Grinning, he added, "Those frustrated 'washouts' will no doubt be looking you up."[21] Later, Knight had dinner with New York mayor Fiorello LaGuardia. The stocky, charismatic LaGuardia was an enthusiastic supporter of the plan, feeling that "the more Americans who could be trained and ready when war came to the United States, the better."[22] He arranged for a meeting with a State Department official, for Knight was still worried about the legal complications of Americans signing up for a foreign war. He had heard too many stories about the problems encountered by U.S. citizens who had joined the French, British, and Italian forces in the last war.

Keeping 'em flying: At No. 6 EFTS, Prince Albert, Sask., in the summer of 1942, civilian ground staff service the unit's Tiger Moths. Most elementary flying schools in the BCATP had been civilian clubs before the war. (John Turnbull collection)

Ground personnel, air force and civilian, kept the BCATP's aircraft flying year-round. At the beginning of the war, the RCAF had only about 1,500 experienced tradesmen; thousands had to be trained without delay. Here, at the Technical Training School, St. Thomas, Ontario, students receive instruction on aero-engines. (DND photo: PL1035)

Statistics: Of the twenty student observers in this class at No. 10 AOS, Chatham, N.B., Sept. 1941, eighteen became casualties. Two were shot down and taken prisoner; one was seriously injured in a crash; fifteen lost their lives.
(Mac Reilley collection)

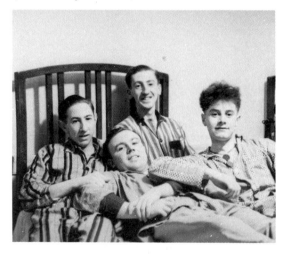

These contented RAF student navigators in Quebec City don't care that a snowstorm rages outside. The weather interrupted their journey to No. 8 AOS, but a kindhearted local arranged accommodation for them at the city's premier hotel, the Chateau Frontenac.
From left: LACs Harrington, Moules, Le Bas, Haynes.
(Doug Harrington collection)

WDs under training as signals officers at No. 3 Wireless School, Winnipeg.
From left: Nell Ross, Mary Gordon, Louise LeClair, Betty Dowler, Dorothy Winter, Patricia Annand.
The aircraft is a Fleet Fort.
(DND photo: PL24469)

Morse instruction at No. 1 Wireless School, Montreal. Many BCATP student wireless operators rated Morse the toughest subject on the curriculum. The first BCATP graduate wireless operators had to take additional Morse training in Europe to ready them for operational conditions there. (DND photo: PL2015)

Armourers at the Air Armament School, Mountain View, Ontario, assemble eleven-pound practice bombs. Filled with titanium tetrachloride, the bombs emitted puffs of smoke on impact, enabling plotters to record how close they were to their targets. (DND photo: PL2062)

Bomb rack: Student observer and armourer check that practice bombs are properly attached to the under-wing rack of a Fairey Battle. Bombing-up could be an unpleasant task in winter. (DND photo: PL3573)

Trainee bomb aimers tested their skill on pyramid-shaped targets floating in lakes. Direct hits on the targets were few and far between, and damage could usually be repaired by replacing some wooden slats. (Canadian Warplane Heritage Museum)

Secure in a bomb plotting tower - a "quadrant" - an airman scrutinizes a bombing range on Lake Ontario, ready to record the results of student air bombers' efforts. (DND photo: PL2095)

Turret time: At No. 10 B&GS, Mount Pleasant, P.E.I., a student air gunner learns to handle the controls of a power-operated turret. Such training equipment as this was in critically short supply in the BCATP's first months. (Norman Thacker collection)

Staff pilot: Sgt. James A. Thomson of Toronto prepares to take off in a Bolingbroke from No. 10 B&GS, Mount Pleasant, P.E.I. Thomson and three students died when Bolingbroke No. 9196 crashed into the sea, April 13, 1945. (Norman Thacker collection)

Aussies: Australian students at No. 1 B&GS, Jarvis, Ontario, ca.1942. Note the shorts and bush hats. The Canadian instructor, F/S Macaleese, stands at the left of the rear row. Another airman seems to be enjoying the proceedings from the nose of the Bolingbroke. (Bert Lee collection)

Aussies on ice: Most BCATP students from down under quickly adapted to Canadian winters. Here, five RAAF student observers at No. 2 AOS, Edmonton, show off their skating skills. Sadly, three of the five - LACs Heatley (left), Wheatley (2nd from right), and McNeill (right) - lost their lives on ops. (Don Charlwood collection)

PR director: The man who did most to make the public aware of the BCATP was a former flier and advertising executive, Joseph W. G. Clark. He energetically pursued the idea of Hollywood making a major movie about the BCATP. The result: Warner Brothers, Captains of the Clouds, starring Jimmy Cagney. (Joseph A. P. Clark collection)

Grand opening: New York mayor Fiorello LaGuardia and Billy Bishop inspect the RCAF honour guard at the premiere of the movie Captains of the Clouds, February 1942. (DND photo: PL6806)

At the conclusion of the 1942 Ottawa Air Training Conference, the highly regarded Canadian air minister, "Chubby" Power (2nd from right), accepts a souvenir from Sir William Glasgow, Australian high commissioner. Left, A. D. P. Heeney, Power's deputy, and Australian A/C V. M. Isaat; extreme right, Alison Sparks, one of a score of WDs working at the conference. (RCAF photo via Alison Tucker [née Sparks] collection)

Needle and thread: WD Fabric Worker A. T. Seaman, of Port Arthur, Ontario, repairs an aircraft's fabric surfaces at No. 6 B&GS, Mountain View, Ontario. (DND photo: PL11430)

Many WDs joined the air force only to find themselves back in the kitchen. At the School of Cookery, Guelph, Ontario, WDs learn to prepare meals of Service proportions. (DND photo: PL6582)

Brace of Bishops: Famous WWI ace Billy Bishop chats with his son, Arthur, after awarding him his pilot's wings at No. 2 SFTS, July 1942. Arthur became a Spitfire pilot and flew a tour of operations with 401 Squadron in Europe. (DND photo: PL9652)

A midair collision demolished two Ansons in February 1943. Remarkably, the three occupants survived the incident. George Stewart, the student pilot of one of the Ansons, graduated from No. 9 SFTS a few days later, receiving his pilot's wings from A/C F. S. McGill. (George Stewart collection)

Gala occasion: Surrounded by Ansons, with plenty of guests and station personnel in attendance, student pilots receive their wings at No. 5 SFTS, Brantford, Ontario. Note the typical BCATP buildings in the background. (Canadian Warplane Heritage Museum)

Full circle: In May 1945, with the European war won, students of the final BCATP class at No. 1 SFTS, Camp Borden, receive their wings. Second from right is G/C J. P. McCarthy, who graduated from the BCATP in 1941 as a sergeant pilot, subsequently flying two tours of ops with Bomber Command. (Joseph P. McCarthy collection)

I was quite certain the mayor had briefed the official on the details of our proposed operation, and wasn't surprised when he was icily and correctly non-committal, and simply quoted the phrasing of the U.S. law which said, among other things: "Hiring or retaining another person to enlist or enter himself in the service of a belligerent as a soldier, or as a marine, or a seaman on board any ship of war. . . ." (It is interesting to note that there was no mention of "airmen" inasmuch as the law dated back to a time before the Wright brothers' contraption had been taken seriously, and incidentally was originally drawn up against the British.) Title 18, Section 21 and 22, U.S. Code, R.S. Section 5282, March 4, 1909, then went on to state the consequences for breaking this law: ". . . he shall be fined not more than $2,000 and imprisoned not more than three years." An additional phrase in an earlier ruling (March 2, 1907) was equally explicit on another point: "That any American citizen shall be deemed to have expatriated himself . . . when he has taken an oath of allegiance to any foreign state."[23]

Meanwhile, Knight noted, Homer Smith and Bishop were "using all their persuasive talents on the military and civilian heads in Canada to abrogate their ruling requiring the volunteers to pledge allegiance to the British King."[24] Their efforts paid off: "Steeped in the oldest traditions of service in time of war, the Canadian military held out against giving up the rule, but finally relented."[25]

The pilots signed up by Smith and Knight had to be "officer material" with at least three hundred hours of certified flying time, a CAA licence, a high school diploma or equivalent, plus a birth certificate showing them to be between twenty and forty-five.[26] (As an added precaution, a confidential report was prepared on each applicant.) They were needed as staff pilots and instructors in Canada and, originally, there was no thought of using them in combat areas. "On that point," said Knight, "we had a little difficulty getting the Canadians to agree. We finally arranged for a clause to be added when they signed up, that they wouldn't leave the North American continent."[27] But soon many of the American pilots wanted the clause removed. Knight recalled talking to an American pilot at No. 2 B&GS, Mossbank, Saskatchewan, who spent his days flying student gunners around. "I didn't come up here to be a damned elevator operator," he told Knight. "I want to fight."[28] Later the pilots sent a petition to the commanding officer of No. 1 Training Command, stating, "We, the undersigned officers of the RCAF feeling that Democracy and the cause of civilization itself is now at stake and that the British Empire alone is defending these ramparts, and whereas, we have enlisted for service on the North American continent, now therefore, we do request that our services be extended to cover *any field of operation* that will best serve the cause."[29]

Everything went smoothly and efficiently for a few months. Then, late in 1940, Mackenzie King's worst fears were realized. The U.S. State Department, prompted, it is believed, by American isolationists, sent a note to Ottawa saying that the Clayton Knight Committee was openly soliciting pilots and had become an embarrassment. Although a continued flow of American pilots was vital if the BCATP was to expand, King had no qualms about closing down Knight's operation immediately. But then came a reprieve. The State Department had apparently acted without the knowledge of the White House. Roosevelt soon put the department officials straight, and Mackenzie King breathed again. The State Department's note had, however, been a sharp reminder of American sensibilities. The Canadian government decided to create a "buffer zone" between the Clayton Knight Committee and the RCAF. Once a candidate had satisfied the examiners in the United States, he would be given expense money and told to report to the Dominion Aeronautical Association in Ottawa. That organization would go through the elaborate charade of rejecting every candidate and then pointing out that the RCAF office right next door "might be interested in your qualifications."

The candidates themselves had little interest in the legal manoeuvrings. They were getting what they wanted. One, Bill Deming, wrote to Knight, saying, "I can't over-emphasize the courtesy and consideration with which we have been treated by all concerned."[30] Another, Vic Mudra, wrote, "I can truthfully say that I've never been treated finer – all expenses paid while waiting, $5.00 a day food allowance drawn daily at the Chateau Laurier's cashier's window."[31] (Knight remarked that the hotel was forced to replace their cashiers several times over; they kept marrying the American pilots and moving away.)

By the beginning of 1941, the Clayton Knight Committee had processed 242 U.S. volunteers. The committee could not indulge in a vigorous advertising program for fear of stirring up reaction among isolationist and pro-German elements. In fact, only one notice was produced, for display in airports across the United States. Slightly smaller than a standard 8 ½ × 11 letter-size sheet, it depicted a Spitfire and carried the message: *If you wish information about joining the Canadian and British air forces or civilian ferry work, write to the nearest information center of the Clayton Knight Committee, New York.* A list of the committee's branch offices was provided.

The committee had since its inception concentrated on recruiting trained pilots, but in the spring of 1941, a new development caused a shift in policy: the flow of Canadian volunteers began to show disturbing signs of drying up. The Air Council, after much heated discussion, asked Smith and Knight to select 2,500 untrained recruits in the United States. Although President Roosevelt had declared a national emergency and had said that U.S. pilots were at liberty to fly in combat with the Allies, everyone from King down was still concerned about the political dangers of recruiting Americans too openly. They changed the name of the committee to the Canadian Aviation Bureau and eliminated all

references to the RCAF on correspondence and forms, substitut-
ing "Canadian Aviation." Ottawa worked out a schedule for the
new recruits, setting quotas for each month. It didn't work. The
young Americans who contacted the Bureau were impatient to get
into action. If the people at Bureau offices didn't take them immediately, they simply headed for the border and signed up at the
nearest recruiting depot in Canada.

When "Chubby" Power became minister of national defence for
air, he lost little time in appointing a director of public relations.
An affable, "red-tape-hating lawyer-parliamentarian," [32] Power
knew that the BCATP needed "selling" to Canadians, who
somehow found it hard to equate the mighty endeavour, the "bil-
lion-dollar university of the air," [33] with military support for hard-
pressed Britain. Sending thousands of young men up to fly
around in little yellow airplanes seemed almost frivolous; the
public expected to see divisions of heavily armed troops marching
up the gangplanks of transport ships. *That* was the sort of military
aid everyone could understand.

Power needed someone to make the public realize the
immense importance of the plan, someone with an intimate
knowledge of the media, and the air force, to get across the story
effectively without heavy-handed didactics. It's hard to imagine
how he could have made a better choice. Joseph W. G. Clark, who
was in his early forties when he took up the post, had won a DFC
as an airman in World War I. A close friend of Billy Bishop, he was
one of the few flight commanders in the RFC who wore the half-
wing of an observer rather than the double wing of a pilot. He
came of a journalistic tradition; his father had for many years been
editor of the Toronto *Star* and his brother Greg was a well-known
columnist with the same paper. In the twenties, Clark joined his
father and brother at the *Star*, after which he went into the advertising business, achieving considerable success and becoming

director of sales for Cockfield Brown, a major agency of the period. He was a dapper, highly literate man possessed of considerable charm – and a singing voice of such quality that a theatrical agent once tried to get him to give up advertising and try his luck in show business. When war broke out, Clark was still recovering from a heart attack suffered during a game of golf. Despite his shaky health, Clark yearned to be involved in the war effort. A telephone call to an old friend, Gaddis Plumb, a former RFC comrade, only intensified his desire. The well-to-do Plumb lived in a mansion on Long Island. When Clark called, Plumb's butler informed him, "Mr. Plumb has departed for England, sir, to join the Royal Air Force." Clark could barely conceal his frustration. Power's invitation to go to Ottawa to do public relations work for the RCAF arrived at just the right time. [34]

To Clark's delight, Power offered him a senior rank in the RCAF. He longed to get back into uniform. However, when he told his brother Greg about his new job, he got a surprise. The older Clark told him it would be a serious mistake to become an officer. It was a trap, according to Greg, who pointed out that as a serving officer Joe couldn't possibly do a first-class job, because he would be hamstrung by the air force structure, with every move having to be approved by level upon level of command. Disappointed though he was by his brother's reaction, Joe saw the logic in his remarks. He told Power that he would only accept the job as a civilian. Power agreed. Initially Clark had a fellow director, I. Norman Smith of the Ottawa *Evening Journal.* The partnership didn't last long, however. Clark and Smith found it hard to agree on anything. Smith resigned, leaving Clark as sole director. When he took up his post, Clark immediately raised air force hackles by insisting that every BCATP airfield be prominently signposted so that people would know where it was and what was being done there. The air force establishment, accustomed to working in semi-secrecy, wasn't happy – and it was even less happy when Clark sent in the hefty bill for the signs.

Within a matter of weeks, Clark had established excellent relations with the press and radio, and soon he became known as *the* source of accurate information about the air force in general and the BCATP in particular. Ignoring his heart condition, Clark plunged enthusiastically into his new job, working long hours seven days a week and travelling extensively.

In the fall of 1940, Clark was negotiating with *Maclean's* magazine on the development of a full-colour feature article describing the work of the BCATP. Clark's son, Joe, Jr., who was about to enter the navy at the time, recalls a lunch at the King Edward Hotel in Toronto at which the magazine editor and his father were present, as well as a former *Maclean's* writer, Norman Reilly Raine. Raine had moved to Hollywood some years before and had become a successful writer with the Warner Brothers studio, responsible for the scripts of such immensely popular films as *Dawn Patrol* and *The Adventures of Robin Hood.*

The men discussed the BCATP and how it could be promoted. Raine's presence undoubtedly planted a seed in Joe Clark, Sr.'s, mind, because he found himself beginning to wonder about the possibility of promoting the training plan with a movie. A *full-length feature,* complete with top stars! It was an exciting thought. He broached the idea to Raine. The writer nodded, interested. The notion had possibilities. Warner might go for it, he thought, adding that Hal Wallis, one of Warner's top producers, was in New York at that very moment. Raine lost no time in calling Wallis, who, perhaps fortuitously, was immobilized in his hotel room, nursing a bad cold. Raine made him promise to stay there until he had seen Clark. Never one to hesitate when opportunity knocked, Clark immediately set off for New York.

The next day Clark found himself in Wallis's suite. A superb salesman, enthusiastic and knowledgeable about his subject, Clark soon had the watery-eyed Wallis sitting up and taking notice. When Clark had finished, Wallis said it would be a great subject for a short. Clark shook his head. No, if the movie was going to do the job, it had to be a feature, with stars and a story. Nothing else would do. Expansively (and without any official backing) Clark promised the RCAF's full co-operation. The moviemakers would be provided with aircraft, airfields, personnel, props, anything that was needed, he declared, hoping against hope that the air force could in fact be persuaded to be that generous.

Despite his cold, Wallis became increasingly enthusiastic. He promised to discuss the idea with Jack Warner as soon as he was back in Hollywood. Meanwhile, Clark returned to Ottawa, probably with his fingers crossed, to tell officialdom what he had been doing. He needn't have worried. Power and his deputy, the former head of Massey Harris, Jim Duncan, nodded their approval. Provided Warner was as enthusiastic as Wallis, there seemed every possibility that the work of the BCATP would be publicized at home and abroad and in the most effective, dynamic way possible. In those pre-television days, there was no better way of doing the job. Thank goodness Raine had been present at that lunch in Toronto. And thank goodness he worked for Warner. Of all the studios then active, Warner had repeatedly demonstrated the most consistent support for the Allied cause, raising funds to purchase Spitfire fighters and producing documentaries about stalwart British resistance during the Blitz, then at its height. Soon Clark had good news from Wallis: the studio wanted to go ahead! In fact, Warner envisaged the proposed film as a major production, a potential box-office blockbuster. No expense would be spared. The BCATP was about to get the full Hollywood treatment.

In January 1941, Clark once again travelled to New York, where he signed contracts on behalf of the RCAF. Wallis signed for Warner. They chose an appropriate venue for the meeting: The Waldorf-Astoria, home of the Clayton Knight Committee.

Although the contracts had now been signed, no script yet existed. Wallis told Clark about a story that had been suggested by the popular Canadian actor Raymond Massey (Vincent's brother), at that time under contract to Warner. Massey urged the studio to develop an outline called *Bush Pilots* written by Arthur Horman and Roland Gillett.[35] Wallis talked about starring Massey and possibly Errol Flynn (Cary Grant was even considered at an early stage) and wanted Raine to write the final script. When the project appeared to be gathering momentum, Dorothy Massey wrote to Wallis: "Ray and I are so pleased that a real, honest Canadian picture is to be made and we wish you great success."[36] Warner paid Horman and Gillett five thousand dollars each for their story outline.

Asked to provide reference materials to help in the development of the script, Clark sent "what amounted to an illustrated encyclopaedia of RCAF material, ranging from a fictionalized student pilot's court-martial, through items such as plotting charts of bombing targets and actual bombing error sheets from No. 1 Bombing and Gunnery School, Jarvis, to cockpit and interior photos of Hudson, Harvard and Fleet Finch aircraft."[37] In February Wallis asked Clark to clear the way for Raine to see the various sites proposed for filming the movie; he also requested that an air force officer be available as technical adviser. A first draft of the script arrived in Ottawa. When Robert Leckie, now an air commodore, read it, he succinctly described it as "tripe" but raised no objections to production going ahead.

Flight Lieutenant Owen Cathcart-Jones, a former Fleet Air Arm pilot and long-distance flier, was working as a liaison officer at the Canadian legation in Washington when the word came from Ottawa that he had been appointed technical adviser and script collaborator on the new, as yet untitled film; "Chubby" Power was not impressed by the original title, *Bush Pilots,* feeling that it didn't convey the "magnitude of our training plan."[38] In mid-April, Cathcart-Jones and Billy Bishop met Wallis and Jack Warner in Hollywood to review recent films, including *International Squadron* (starring Ronald Reagan), *Dive Bomber,* and *I*

Wanted Wings, to point out errors in dress and deportment of various actors portraying RAF airmen. Jack Warner promised that no such errors would mar his new movie. It would be accurate in every detail and, he declared with typical Hollywood modesty, the biggest production of the year – a "boffo" hit. By now the studio had decided not to star Massey, Flynn, or Grant. Moustachioed, slick-haired George Brent had been picked for the starring role. An urbane actor who usually played slightly anaemic leads in romantic films, Brent seemed to some people an odd choice to portray a Canadian bush pilot. But actors were Warner's business, not Ottawa's. Clark and his colleagues had enough to do making arrangements for sites in the lake country near North Bay, Ontario, and at BCATP fields at Ottawa, Trenton, and Jarvis.

Another scriptwriter, Richard Macaulay, came in to assist Raine. The plot now involved bush pilots working in the northern wilderness who volunteered to serve in the RCAF on the outbreak of war. The drama revolved around the conflict between the two principal characters: the upright, honest-as-the-day-is-long bush pilot and the brilliant but cocky one, who never plays by the rules and who seems at first to have some questionable qualities, but in the end proves himself a hero beyond compare. In that era no aviation epic was complete without such a character. The brash, boastful persona seemed to strike a chord with American males, a reflection, perhaps, of the way they saw themselves: as rugged individualists contemptuous of rules and regulations, interested only in getting the job done while winning the heart of any female who happened to be in the vicinity. As Jerry Wald of Warner told Hal Wallis, "audiences always like a cocky guy who boasts about his ability and then comes through."[39]

What was the film to be called? Between them, Warner and Clark came up with dozens of title suggestions, including *Shadow of Their Wings, The Fighting RCAF, Wings of Canada, Bush Pilots Over the Sea, Crusaders of the Air.* In the end Warner picked a phrase from a speech by Billy Bishop in which he had appealed for contributions to a Victory Loan drive and referred to Common-wealth airmen as "captains of the clouds." Everyone loved it.

Before production began, Warner announced that Brent had been replaced by James Cagney. There is no doubt that Cagney was the right actor to play the part of Brian MacLean, the rebel with the stout heart of pure gold. The script described him as "restless, mercurial . . . but he has unsuspected depths of emotion and a capacity for feeling which only the sharp wrench of tragedy will bring forth." [40] The trouble was, Cagney had been playing the part for years, as pilots, cowboys, racing drivers, and gangsters. When he read the script he didn't like it; it was just the same old thing dressed up in RCAF blue. Jack Warner had to convince him that he would be helping the Allied war effort by accepting the part. Only after much discussion – and Warner's agreement to take on William Cagney, James's brother, as associate producer – did Cagney agree to play his stock character yet again. Dennis Morgan, an amiable actor with a pleasant smile and curly hair, played Johnny Dutton, Cagney's rival and foil. Alan Hale, the bluff and not very bright Little John of *Robin Hood,* played essentially the same role in *Captains of the Clouds,* portraying Tiny Murphy, the bluff and not very bright bush pilot, too old for aerobatics but too stubborn to admit it, even to himself. Brenda Marshall, a dark-eyed beauty who had created something of a stir playing opposite Errol Flynn in *Sea Hawk,* had the unrewarding part of Emily Foster, Johnny Dutton's girlfriend, who, inevitably, falls for MacLean. Interestingly, in an early draft of the script her name was Emilie Fradette, and she was described as "a roving-eyed little dame who plays around with every guy who comes along." [41] Each succeeding draft blunted the edges of her character and elim-inated her French connections. Originally the studio considered a string of actresses for the role, including Ida Lupino, Betty Field, Rita Hayworth, and Fay Emerson. Marshall became the official choice only a few days before production began in the summer of 1941. In early drafts of the script, there was an assembly of names for the pilots and other characters in the Canadian bush that sounded as if they had sprung from the pen of Damon Runyon; for example, First Class Freddie, Pincushion Kelly, One-No-Trump O'Connor, Soup-Tureen Murphy, Nickle-A-Schnickle Blomberg. [42] Mercifully none survived into the shooting script. A director named Keighley was the first choice for the film, but Michael Curtiz eventually got the job, although, according to Warner production notes, the company was concerned because Curtiz seemed to want to make an epic of the story – "a *Gone With the Wind* of the air." [43] (A year later Curtiz would win an Oscar for his work on *Casablanca*.)

Production began. The Warner contingent, numbering about eighty, immediately ran into problems at the border. Immigration authorities suspected Sol Polito, the chief cameraman, of being an enemy alien. Only after a tiresomely long investigation was Polito permitted to enter Canada, in company with the crew of electricians, grips, wardrobe and makeup people, and all the other specialists needed for a location shoot. They reached Ottawa . . . and promptly threatened to strike. They didn't like the tents that had been set up for them at No. 2 SFTS; they wanted solid roofs over their heads. After much scurrying around by RCAF personnel, accommodation was arranged at the Sergeants' Mess and other buildings at Uplands. So far so good. But not for long. A lack of soap and towels almost created another crisis. More scurrying by the air force to meet these demands. Meanwhile, the stars and executives had been booked into the Chateau Laurier hotel, which provided soap and towels in abundance.

Poor weather washed out filming for several days. When at last the weather co-operated, the governor general, the Earl of Athlone, decided to visit the set, and production again ground to a frustrating halt. The distinguished guests said polite things the stars and crew before moving on. Leckie's reaction to the delays in training because of the "tripe" being filmed can be imagined.

Billy Bishop portrayed himself in the movie, presenting wings to successful graduates at Uplands. He revelled in the whole business, according to his son, Arthur, reeling off his lines without a hitch. But the daily progress reports to the studio told of a series of problems at the shoot: "Air Marshal Bishop wrote his own speech which ran for five minutes, and was OK'd by Mr. Wallis, instead of a few lines as in the script. This conservatively added two days to Curtiz's schedule while at Uplands."[44] In Hollywood, the studio kept complaining about the rising costs, while the unit in Canada kept pointing out that they had been sent up north without adequate preparation and that everything from uniform material (seven dollars per yard) and buttons (small: twenty-eight dollars per gross, large: thirty-six dollars per gross) to aircraft servicing cost much more than it did in California.[45]

The script called for high-spirited bush pilots to buzz Uplands during a ceremony at which Bishop was presenting wings to graduating pilots from the Commonwealth and the United States, disrupting proceedings and incurring the wrath of the air force. "My father," writes Bishop, "standing rigidly to attention, was to appear startled and annoyed, yet to maintain military aplomb, as he watched the two aircraft dive down, sweep across the field and climb away. . . . The trouble was there were no aircraft to look at – only a stagehand holding a red flag running along an elevated ramp in front of my father at a height slightly above eye level. The man with the flag sometimes ran too fast, sometimes not fast enough and once he nearly fell off the ramp. The scene had to be shot over and over, until at the end of the warm day the exasperated Curtiz exclaimed, 'This is the last take positively if I have to take an aeroplane up myself.'"[46] Arthur Bishop recalls Curtiz becoming increasingly frustrated about the miserable weather which had descended upon the Ottawa region as soon as the film crew arrived. With his production schedule in disarray, Curtiz told his associates, "Hot or cold, tomorrow we go!"[47] And go they did.

It was hard work, although no one but the actors and crew knew it. In his autobiography Cagney recalled how much energy the principals – Alan Hale, Dennis Morgan, and he – expended doing one scene. The action seemed simple. They were to bring in a plane, jump out of the cockpit, then sprint fifty or sixty feet to get out of camera range. It took all day. On one take the aircraft was in the wrong position, on another the actors were in the wrong position; the sun disappeared; the onlookers weren't facing the right way. It went on and on. At day's end, the three weary actors returned to the Chateau Laurier to relax. There was a knock at the door. Cagney opened it to find a young man who wanted to see Alan Hale. Thinking the young man knew Hale, Cagney invited him in. Hale and Morgan assumed that the newcomer was a friend of Cagney's. Although it quickly became apparent that no one knew the young man, he made himself at home, asking the actors if movie-making involved any work, or was it all fun?[48]

In Ottawa, the Hollywood invasion was the most exciting event since the Royal Tour of two years before; crowds of fans surrounded the stars whenever they ventured out into the city.

The air force pilots who flew in the movie had their own problems. Tom Anderson was an instructor at Uplands at the time and was recruited to do some of the formation flying in Harvards. He recalls that the director wanted the aircraft with a good background of cumulus cloud. This meant that the formation had to go higher and higher to find just the right conditions. In the end the shot was taken at about fourteen thousand feet. At that frigid altitude, without oxygen, Anderson caught a bad case of the flu and spent a week in hospital recovering.

In typical movie fashion, the beginning of the film was shot last. The crew and stars moved to North Bay, where a fictional trading post, Lac Vert, had been built some forty miles from the city on Trout Lake's Four Mile Bay. Again the weather failed to co-operate. Rain fell day after day. Lightning damaged the set. The leaves began to take on their fall colours, creating continuity problems.

As if that wasn't enough, a husky dog bit Dennis Morgan. Cagney suffered a mild concussion when he hit his head during a scene. Industrious beavers flooded the road leading to the site and some vehicles became mired. One slid down a ravine, injuring two occupants. The players staying at the Empire Hotel, North Bay, found it hard to rest because of the wide-eyed fans constantly swarming through the hotel, hoping to glimpse, or better still meet, them. Eric Stacey, the unit manager, complained of the "swarms of tourists and kids that simply infest the lobby of our hotel,"[49] and he censured the crew for bringing women into the hotel. Cameron Shipp, Warner's publicity man on location, had to write scores of regret notes to such groups as the Kinsmen Club of North Bay, who kept inviting Curtiz, Cagney, Hale, Morgan, *et al.,* to dine with them, and to individuals like the lady in Ottawa who wanted Curtiz to come over for an evening and view her scrapbook of the Royal Tour. "Nothing would have been more delightful, but time simply won't permit . . ." went Shipp's letters. He had also to write to dozens of individuals who wanted parts in the film. One was a pilot at Trenton; another a girl from Niagara-on-the-Lake, Ontario, who described herself as a "young tomboy who craves adventure and excitement. . . . Send for me; you won't be disappointed."[50] But the hopes of such aspiring thespians were to be dashed; the only people hired locally were Indians. Stacey got the Hudson's Bay Company to find "Six good Indian types – men." He wanted one old man – "well wrinkled" – plus two of about forty years of age, and three young men, from twenty to thirty. He also wanted some women: one old and, of course, "well wrinkled," one "middle-aged," and three youngsters. The film company paid the Indians three dollars per day, plus lunch, for sitting around and looking authentic.[51]

Paul Mantz leased the company his three-engined Stinson Model "A" for air to air photography, all of it in Technicolor. The pilots engaged by Warner for the flying sequences were Jerry Phillips, Frank Clarke, Howard Batt, Harry Crosby, Richard Rinaldi,

Paul Gustine, and Frank Tomick. They had trouble with the glassy-calm conditions prevailing on Trout Lake for most of the shooting. A Norseman rented from Dominion Skyways was damaged in a heavy landing, and the subsequent claims strained relations between Warner and Dominion. Doubling for Cagney, Jerry Phillips flew the Norseman on many occasions. One particular sequence "called for me to land on the lake in front of the pier, then taxi up to the pier, cut the engine and let the plane coast the last few feet into the pier. In the meantime I was to duck down out of sight and Jimmy [Cagney] was to climb out onto the float nearest the pier and throw a rope to Brenda Marshall who was on the pier. During the rehearsal the director wanted Brenda to show anger because she was supposed to be mad at Jimmy for two-timing her. In his broken English, Mike Curtiz said, "Brenda, I want for you to be mad at Jimmy like the bitch you are!" Brenda Marshall took it well, Phillips said.[52]

Eventually, to everyone's intense relief, the filming in Canada came to an end and the crew returned to Hollywood for interior shooting and editing. At Cathcart-Jones's urging, Bishop accompanied Clark to the motion picture capital for the final cutting in November. Bishop enjoyed the hospitality of the movie people so much that he stayed far longer than he had intended, to the considerable annoyance of his wife in Ottawa.

Bishop and Clark were delighted with the final product, both men sure that it would realize its goal of generating interest in the RCAF and stimulating recruiting.

Despite Leckie's opinion of the script, *Captains of the Clouds* proved to be immensely popular with the public. It was a hit, "a stratospheric flyer for outstanding box-office attention," according to *Variety*'s febrile prose. "Timely, topical, and strongly patriotic in theme, it zooms along at a zestful, attention-arresting pace to rank as a holdover of no small pretensions for all runs." The show business journal pointed out in its inimitable way that the film was a "particular patriotism-rouser for audiences in all parts of the British Empire."[53] The *New York Post* praised the "changing forcefulness" of Cagney (whatever that meant), and pointed out that no one in the film matched him "except, oddly, a stoutish man who is not a professional actor at all. Air Marshall [*sic*] Billy Bishop, Canada's greatest flying ace in the last war, makes a speech to airmen who are about to receive their wings. It has a ring of iron to it, although the manner is not forced."[54] The New York *Herald-Telegram* was ecstatic: "*Clouds* literally roars with excitement. The flying scenes are breath-taking and will leave you limp with suspense."[55]

For Joseph Clark, all the effort had been worthwhile. The film publicized the BCATP far more effectively than a thousand magazine articles. Ironically, though, it did little to encourage Americans to join the RCAF, for while Warner was getting *Captains* ready for release, the Japanese bombed Pearl Harbor. Everything changed. Overnight, young Americans wanted to join their own services, not Canada's. Many Americans already serving in the RCAF wanted to transfer without delay. The work of the Canadian Aviation Bureau, a.k.a. the Clayton Knight Committee, came to an abrupt standstill. The organization would soon be out of business, its day done.

Today *Captains of the Clouds* is remembered as just one of countless war movies, epics in which our guys are always gallant, charming, and witty, while the enemy soldiers/sailors/airmen are vicious and merciless and possess not a glimmer of a sense of humour. Now *Captains* seems rather silly, its characters one-dimensional, its story line trite and predictable. As a romance it fails miserably; the stars evince little passion for one another (despite *Variety*'s assessment of the romance as "lusty"). While the film is better than most from a technical point of view, the flying sequences so lavishly praised in 1942 can be divided into two categories: the genuine aerial film scenes, which are of more than a little interest, and the fakes, which are laughable. Again and again Warner used not-very-convincing model airplanes that darted and lurched around the screen, looking as much like the real thing as images in a computer game. (Jack Warner later complained about the cost of these sequences, adding ominously that in his opinion the miniatures department was 50 per cent over-manned.) The most appealing parts of the movie are the scenes of the BCATP in action, which provide a compelling glimpse of another time, a chapter in Canadian history of half a century ago.

CHAPTER EIGHT

The Road to Maturity

Westland Lysander

"... this second Commonwealth air training conference had moved the RCAF some way down the road to maturity and independence"
— *Official History of the RCAF,* Vol. II, p. 263

It was a time of good news and bad. On one hand there was the immensely encouraging fact that the Americans had at last come into the fight; on the other, the unwelcome addition of a new enemy that had already shown itself to be infinitely tougher and more capable than anyone had anticipated. In terms of numbers and productive capacity, there could be no doubt that the Allies held all the advantages. On paper, the Axis powers didn't stand a chance. But battles aren't fought on paper. In a matter of months, the Germans had occupied a staggering half-million square miles of the Soviet Union. No one questioned the courage with which the Russians defended their country. But how long could they hold out? Total defeat of the Soviets loomed as a very real possibility, with heaven only knew what horrific ramifications. In the Middle East, after a successful offensive in Libya, the British were falling back in the face of Rommel's brilliantly planned and executed counteroffensive. The Far East was an equally dismal picture early in that new and infinitely depressing year of 1942. Hong Kong, Singapore, and Bataan had fallen in appallingly rapid succession, stunning those Americans and Europeans who had been so dismissive of the Japanese. Short-sighted soldiers, flimsy airplanes that were copies of outdated European and U.S. designs, pilots who couldn't take G-forces, sailors who got seasick in storms – such were the myths widely and confidently circulated in the days before Pearl Harbor. Not only in the Far East did the Axis powers humiliate the Allies. In the English Channel, on Britain's doorstep, the German battleships *Scharnhorst* and *Gneisenau* and the heavy cruiser *Prinz Eugen* cocked a contemptuous snoot at the might of the Royal Navy and RAF as they made a daring escape from Brest. An incredible tonnage of British and American capital ships lay rusting on ocean floors all over the world. Could the Axis powers actually *win*?

Although no one knew it, the tide was beginning in its unhurried way to turn. Midway. El Alamein. Stalingrad. Unfamiliar at first, the names soon became synonymous with victory, irrefutable proof that the Allies were on the right road. The change in Allied fortunes would affect every aspect of the war effort, including the BCATP in peaceful Canada.

For some time prior to the United States' entry into the war, Ottawa had been attempting to organize a conference at which all interested parties could discuss the future of the BCATP. Such a conference was needed, Ottawa felt, because the original agreement had little more than a year left to run. The war situation had changed out of all recognition. How should the plan be modified to reflect these changes? In August 1941, Churchill and Roosevelt had met at sea off Newfoundland to draw up the Atlantic Charter. To most people in the Allied camp, the meeting was a hopeful sign of closer ties between Britain and the United States. To the touchy Mackenzie King it was nothing short of an insult. He had long considered himself the indispensable link between Britain and America, a key player in the lofty deliberations that would determine the fate of much of the world. How could the two leaders meet off Canada's coast without inviting Canada's prime minister to join them? And not even let him know that the meeting was taking place? Seething, he criticized Churchill and Roosevelt for running such dreadful risks to attend the meeting at Placentia Bay – "a gambler's risk... the apotheosis of the craze for publicity... a matter of vanity." [1] Not only did the president and prime minister put their own lives at risk, he complained, they had endangered many of their ministers and chiefs of staff as well. He soon calmed down. When he flew to London for a meeting with Churchill later that month, he made no mention of Placentia Bay. For all his brash confidence at home, King often seemed ill at ease in the presence of the world's mightiest. It was a characteristic he shared with many of his colleagues. "The Canadian ministers were colonial, provincial politicians, largely ignorant of the personalities and ways of world politics, and, once outside their own limited local experience, easily intimidated and imposed upon. . . . Vincent Massey, the Canadian High Commissioner, a man of the world in his own right and deeply suspected by King for that very reason, noticed with surprise and regret that the Canadians [in London] 'frequently behaved like awkward and tongue-tied country cousins.'" [2]

King probably regretted not giving vent to his feelings about Placentia Bay, for shortly after the attack on Pearl Harbor, London and Washington worked out a deal concerning the allocation of aircraft. Now the Anglo-American Munitions Assignment Board controlled the entire process, and Ottawa hadn't been involved in those discussions either. In fact, Ottawa hadn't even been informed that they were taking place. Colonial attitudes died hard. Adding to the Canadians' discomfiture, the Americans summarily banned the export of aircraft, engines, and spares until further notice. In Ottawa, politicians and airmen wondered and worried. What effect might this have on the BCATP?

The normally affable "Chubby" Power became increasingly irritated. Didn't the Americans understand that if they maintained their stranglehold on aircraft and parts, the entire Air Training Plan in Canada would be imperilled? Shortages of operational aircraft in Britain meant that trained pilots and other aircrew were arriving from Canada and sitting around with nothing to do. The Air Ministry might soon be asking for the output of the BCATP to be *cut back*, incredible as that might seem after all the efforts to increase the supply of aircrew.

The availability of aircraft and engines had long been a source of concern to senior RCAF officers. Back in April of 1941, Air Vice-Marshal E. W. Stedman, the air member for aeronautical engineering, had written to Power pointing out that the future of the BCATP could be jeopardized if early decisions were not made about future requirements for aircraft. But such decisions could not be made without knowing the plans of all nations concerned. Thus a conference was essential. After it had taken place, orders could be placed in good time to permit manufacturers to prepare their production lines. "Having heard that American factories were fully committed to June 1943, he [Stedman] feared that if orders for aircraft were not placed in good time the BCATP would find itself with insufficient training machines. Canada could not place orders without consultation with her partners," writes historian W. A. B. Douglas.[3]

The British agreed to a conference. But they wanted it in London. The Canadians said no. To them, the BCATP was the "symbol of Canada's wartime achievement."[4] It had been created in Canada and it was run by Canadians. And the biggest part of the cost was being borne by Canadians. If there was to be a meeting of the countries involved, it must be held in Ottawa.

London didn't respond. Perhaps this shouldn't have surprised the Canadians. Although the BCATP was a major fact of life in tranquil Canada, in a London reeling under air raids and a deluge of dismal news from every war front, it was hardly the stuff of headlines. Not until January 20, 1942, was there a response to Ottawa's note, and it proved disappointing. London suggested that the conference should be postponed "until the effect on Allied air training requirements of Japanese and American entry into the war could be assessed."[5] Although the message did nothing to solve the Air Training Plan's problems, it was hard to disagree with its logic. The rapidly deteriorating situation in the Far East might necessitate major changes in the plan. What if Australia and New Zealand decided to withdraw all their students and bring them home to help defend their countries against a Japanese invasion? Moreover, what was the point of having a conference about aircraft and engine supplies when the Anglo-American Munitions Assignment Board now exercised total control over their allocation? To the further dismay of the Canadians, the British then suggested having an Empire Training Conference – with American "observers" present – in London at the end of January. Complicating the whole business was the fact that late in December the American representatives to the Permanent Joint Board on Defence (PJBD) had alarmed the Canadians by proposing "that the Canadian and United States Governments should consider the advisability of arranging for a meeting of appropriate representatives of Great Britain, Canada and the United States to make appropriate recommendations for coordination of the entire aviation training programs to be conducted in Canada and the United States."[6] The suggestion alarmed Ottawa's air force hierarchy. What were the Americans up to? Did they intend to take control of training in Canada? Were they out to swallow up Canada's pride and joy, the BCATP, in their own training organization? The Americans denied any such intention, saying simply that they thought it worthwhile to discuss whether North American training facilities were being put to the best possible use.

Perhaps, they added, the RCAF might like to utilize bases in the eastern states during the winter months when flying training was often curtailed in Canada due to bad weather. Shying away, as if believing the proposal to be some form of devilishly clever trap, Lloyd Breadner, by now the Canadian chief of the air staff, refused even to discuss the matter.

Canadian hackles rose higher when Ottawa learned in February that London and Washington had been discussing the training of Fleet Air Arm pilots in the United States – and once again they hadn't troubled to involve Ottawa in the talks. At the same time, General H. H. Arnold, chief of the U.S. Army Air Forces, added fuel to the fire by proposing a meeting of American and British "aviation training people" to review all the training facilities in North America. It appeared that the Americans had every intention of absorbing the BCATP, if they could get away with it. And the British were aiding and abetting them! Power's mood darkened further. He wanted the Americans at the Ottawa conference, if one could be organized, so that these thorny matters could be addressed openly and frankly. He also wanted to impress the United States representatives with the size and importance of the BCATP in order to force an early resolution of the aircraft supply problem. The British suggested a two-part conference, the first dealing with air training problems in general, the second dealing only with Commonwealth matters.

The weeks dragged by while the news from the war fronts became progressively bleaker. The Japanese entered Rangoon. Rommel pushed the British back across Cyrenaica, retrieving all the territory Germany had lost a few weeks before. In Russia, the Soviets' short-lived winter offensive ground to a halt; the Germans readied themselves for a gigantic, multi-pronged offensive that would, they believed, put paid to the Communist state once and for all.

In March, Clement Attlee, the British deputy prime minister (and leader of the postwar Labour government), talked to King about transferring more RAF schools to Canada, adding encouragingly that opinion in London seemed to be shifting to the view that all RAF schools in Canada should come under BCATP control. It was a brave effort, but it didn't calm Ottawa. Canadian-British relations became distinctly strained, at least at the diplomatic level, one side being "not at all satisfied" while the other talked about "unfortunate" attitudes. Canada, as ever the small power squeezed in between two giants, did its traditional best to influence both sides. In this instance, the country's position was particularly difficult. While Canada wanted the Americans to understand that the BCATP was not an organization to be taken over like some minor branch plant, the Plan was utterly dependent upon an assured supply of American aircraft and engines. Ottawa feared that the British proposal to exclude the United States from the second part of the conference might suggest that the Commonwealth nations had organized themselves and had worked out their positions ahead of time. Although uneasy about the United States' ambitions, the Canadians had no wish to offend their friendly but overwhelmingly more powerful neighbour to the south.

In mid-April the prime minister visited Washington. Placentia Bay notwithstanding, Mackenzie King admired Roosevelt and revelled in sharing the spotlight with him, talking to him man-to-man, leader-to-leader. He delighted in the fact that Roosevelt called him "Mackenzie" – and, although a non-smoker, King gladly put up with FDR's incessant puffing of Camels. The two leaders discussed a variety of subjects, one being the proposed conference on air training. Roosevelt surprised King. "The President at once said he thought that was a good idea," King noted in his diary, "and I then suggested possibly South Africa in addition to Australia and New Zealand. The President then spoke of . . . Norway, the Netherlands and China."[7]

The civil servants in Ottawa blinked in disbelief. They had spent months sending messages, trying to drum up interest in the conference, and getting nowhere. Now everyone wanted the conference – and wanted everyone to attend. King and Roosevelt collaborated on a press release:

> The Prime Minister of Canada and the President announced today that, at the invitation of the Prime Minister, a conference in which all of the United Nations with Air Training Programmes would be invited to participate would be held in Ottawa early in May . . . the Conference developed out of the recognition of the desirability of more closely co-ordinating the British Commonwealth . . . Air Training Plan with the greatly extended Air Training Programmes undertaken by the United States and the other . . . United Nations . . . these will include China, Norway, the Netherlands and several others . . . already at war with the Axis. [8]

The sudden snowballing of the conference invitation list created special problems. Invitations would have to go to "Poland, Czechoslovakia, South Africa, Yugoslavia, India, and the USSR, as well as to those nations mentioned in the communique. In addition, Washington was to be asked to recommend which of the Latin American countries should attend, while London would provide information on 'which of the other United Nations have embarked on air training programs.' Eventually the Free French and Belgians were added to the list because they were represented in the RAF schools." [9] But in the case of Greece, the officials were in a quandary. Did Greece have an air force? No one seemed to know. Few of these smaller nations ran air training schemes; they simply supplied students for others to train. Ottawa felt that such countries should be satisfied with observer status at the conference, but not all of them agreed. The Norwegians pointed out that they and other occupied countries could not provide a steady flow of aircrew candidates indefinitely. At some point they would have to tap their only remaining manpower source: nationals living in Canada, the United States, and other Allied nations. It was another topic to be discussed at the forthcoming conference, another sub-committee to be formed, to join those on aircraft production, training methods, co-ordination of training, and so on.

Daily the conference agenda grew larger and more complex. The Canadians, who as hosts were responsible for the agenda, felt deeply about only three subjects: first and foremost, the supply of aircraft and engines for the BCATP after March 1943; second, the persistent reports that aircrew courses might have to be lengthened in order to reduce the surplus of trained airmen overseas; and third, the possible expansion of Canadian training facilities to meet the needs of the British (and, perhaps, those of the Americans too, for they were still talking about co-ordinating North American training facilities). They were topics that "could not be discussed in a vacuum. The flow of students through the BCATP would depend ultimately on the availability of operational aircraft and the need for aircrew overseas, as well as the supply of training machines."[10] The BCATP was a vast organism, nourished by a constant supply of trainees, depending for its existence upon a complex infrastructure of *matériel* and personnel, born of the needs of great air forces fighting on many fronts. Clearly, a full and frank discussion of plans and problems was essential. Unfortunately, although understandably, the Americans and British were reluctant to discuss their strategic plans before too broad an audience. The conference that had been so difficult to bring into being had now become too big for its own good.

The Ottawa Air Training Conference began on May 19, 1942. Mackenzie King welcomed the delegates and talked of the value of the experience gained in air training since the BCATP had become a reality more than two years earlier. Harold Balfour, Billy Bishop's old comrade from World War I and Britain's representative at the conference, pleased King and the other Canadians by describing the country as "the hub of air training around which revolves the ever-widening circle of world battle."[11] Robert Lovett, the U.S. representative, also did his bit to salve Canada's bruised pride by referring to the nation as "the Aerodrome of Democracy,"[12] a phrase which was to acquire an almost Churchillian immortality.

The first order of business was to elect Power chairman of the conference. He also became head of the committee responsible for co-ordinating the work of the conference. Baron Silvercruys of Belgium was in charge of the general training committee; Air Marshal A. G. R. Garrod, the British air member for training, headed the committee on the standardization of training; Robert Leckie, now an air vice-marshal, was in charge of the committee on the composition of aircrew. Two Americans, Captain A. W. Radford, director of training in the U.S. Navy Bureau of Aeronautics, and Colonel R. E. Nugent of the War Department's general staff, took charge of committees dealing with training capacity and manpower resources. Matters of broad interest to all delegates were to be discussed first, with the BCATP discussions coming later – a reversal of the format originally proposed by London.

The arrival of so many high-level delegates posed a major security problem. Scarlet-uniformed RCMP officers stood guard at every door at the temporary buildings erected on Sussex Street (later renamed Sussex Drive, on which the prime minister's official residence is located) to house the conference headquarters, demanding ID from everyone, no matter how exalted their rank. Alison Sparks (now Tucker) was one of the WDs working there, assisting Squadron Leader P. A. Cumyn, the conference secretary. Much of her time was spent delivering documents from the conference to the embassies of the nations involved. Usually she travelled with an RCMP escort. "The WDs worked long hours," she recalls, citing one day when she was on the job for twenty-four hours "with no sleep at all." The conference had its lighter moments, however. "While waiting for one conference to end," says Tucker, "I decided to try on one of the hats left by an air vice-marshal. It was adorned with lots of gold braid.... Unfortunately, I did not hear AVM [G. O.] Johnson opening the door and was caught in the act of marching back and forth with his hat wobbling precariously." Johnson took it well, thanking Tucker for looking after his hat while he was busy.

The conference saw yet another attempt to co-ordinate the training carried out by the RCAF/RAF and the United States army and navy. It was doomed from the start, each side being convinced that *its* approach was right. To each, "co-ordination" meant the other fellow giving in. The solution was typically diplomatic: the formation of a standing committee to work toward further standardization of training techniques among the Allies. The Combined Committee on Air Training was announced in June 1942. After months of delay, it met in April 1943, nearly a year later. It met seven more times during the war, alternating between Canada and the United States. It is unclear what it accomplished; for the balance of World War II, the training programs in Canada, Britain, and the United States followed their separate paths as they had always done. Acres of correspondence passed between the members of the committee, all of it presumably read and digested – then filed away. Perhaps it was enough that the committee *existed*; the politicians could, with some justification, say that they had done everything possible to ensure co-ordination of training among the Allied powers. The fact that the committee's efforts had resulted in little if any action could always be blamed on someone else.

On May 22, the sessions dealing with the BCATP began. No one disagreed that the plan should be extended beyond March 1943. But how far into the future should it run? With almost uncanny prescience, the delegates decided on March 31, 1945. They missed the end of the war in Europe by little more than a month.

Two developments seemed likely to have the biggest impact on the BCATP: the shortage of operational aircraft overseas and the changing composition of RAF bomber crews. There was concern that either or both of these factors could necessitate a reduction of aircrew output. However, it soon emerged that the shortage of operational aircraft was only temporary. In a matter of months, the BCATP would be busier than ever, the Canadians were assured. Vast air forces were being planned for the offensives that would eventually defeat the Axis powers; more schools and more instructors were going to be required to produce aircrew in huge numbers. And all of these expanded training facilities would come under Canadian control.

The conference of May 1942 turned out to be a watershed in the history of the BCATP. While it was being organized, the Canadians had been dubious about the intentions of their allies and had worried that the BCATP would not remain under their control. After the conference they found themselves in complete charge of a greatly extended program. Not that the British were totally happy about this; the old comments about the RCAF having little administrative depth came out for another airing, but in the end there was agreement.

The BCATP was Canada's to run.

Arthur Harris had taken over RAF Bomber Command in February 1942 and had immediately started the long job of building up his force. He envisioned a mighty armada of heavy bombers – as many as four thousand. Such a force, he felt, would be capable of utterly destroying Germany's industrial capacity and, inevitably, the country's ability to wage war. Moreover, the job could be done with little or no help from the army or navy. It was an amazingly optimistic claim from the chief of a force that had up to that time been one of Britain's major disappointments. While the public believed that most of Germany's big cities already lay in ruins, the nation's social structure wobbling under the strain, the War Cabinet knew better. Bomber Command's youthful crews kept trying and kept dying, but their efforts weren't hurting Germany. Most of their bombs fell in open country or on the wrong targets. Many senior officers thought the force should be disbanded, its aircraft diverted to combat the U-boats, at that time a particularly serious menace.

Harris recognized the necessity of convincing the government – and the public – of Bomber Command's true power and potential. He did it by means of the famous thousand-bomber raids in the spring of 1942 that created headlines around the world. Harris had experienced the greatest difficulty assembling a force of a thousand bombers; at the time, his frontline strength numbered about four hundred aircraft. Originally he counted on Coastal Command making up the difference – after all, most of their heavy aircraft had been "borrowed" from Bomber Command in the first place. But the Admiralty said no; the aircraft were needed to fight the U-boats and couldn't be risked. Harris had no choice but to dig deeply into his manpower reserves, using OTU aircraft flown by instructors and crews in the latter stages of training. Frank Hamilton of Mazenod, Saskatchewan, took part in all three of the thousand-bomber raids. He had recently graduated from the BCATP in Canada and had completed most of his training at No. 12 OTU, Pershore, Worcestershire. In the normal course of events, he would have gone to a squadron and flown on one or more ops as second pilot with an experienced crew before being sent off as a captain. But Harris wanted every available crew to be in on the big raids. Hamilton flew as captain in command of a Wellington with a crew of fellow trainees. After the raids, Hamilton recalls, the crew members were informed that they had graduated.

When Harris took over the bomber force, the big new four-engined aircraft – the Stirling, the Halifax, and the Lancaster – were coming into service in significant numbers, although the twin-engined Wellington, Hampden, and Whitley still equipped most squadrons. The Manchester, of which so much had been expected, was a failure, its massive twenty-four-cylinder Vulture engines proving unreliable, never delivering the power Rolls-Royce had promised, and exhibiting a catastrophic proclivity for bursting into flames at inopportune moments. After only a few months in service, the unpopular Manchesters were already scheduled for replacement. Most squadrons flying them received Lancasters, essentially the same airframe with four Merlins replacing the treacherous Vultures.

The size and complexity of the new generation of bombers had a considerable impact on the BCATP even before the May conference. Up to that time, large aircraft had two pilots, a captain and a "second dickey," usually a junior pilot acquiring experience before being assigned his own crew. Although the United States favoured the use of two pilots in large aircraft throughout the war, the RAF came to the conclusion that the practice was wasteful and unnecessary. Of all aircrew, pilots took the longest and cost the most to train. Why not use only one in each heavy bomber, with some other member of the crew assisting him when necessary, which in the normal course of events meant during landing and take-off? Additional duties, such as switching fuel tanks and monitoring the aircraft's systems, could be handled by someone else. It didn't take a trained pilot to pull levers and press buttons. No one pretended that the new arrangement was ideal, but with the enormous expansion of Bomber Command already in the works, it had to be tried.

The job of the observer came under scrutiny too. His principal responsibility, navigation, had become a lot more demanding. The era of the "bomber stream" had arrived. Masses of bombers flew close together, accepting the risk of collisions in exchange for the benefits of getting the bombers through the target area as rapidly as possible, "swamping" defences with their immense numbers. Extremely accurate navigation was required of every aircraft; for the navigators in their vibrating, curtained-off compartments, it was a nonstop task from take-off to landing. Another factor was the assortment of navigational "black boxes" that began to occupy the last available nooks and crannies of Harris's bombers. Each required someone's attention. It was all more than one man could handle. Thus the observer's badge, the winged "O", was replaced by a navigator's badge, much to the chagrin of many veteran observers. (For some types of duties and for lighter bombers and intruders, specialist navigators were trained; for example, navigators "B", who received training in bomb-aiming, and were in most cases assigned to maritime operations, and navigators "W", who combined navigation with wireless skills.) A new aircrew trade appeared, that of air bomber, usually referred to by fellow airmen as the bomb aimer. (As the impact of Harris's raids grew progressively more frightful, many bomb aimers became leery of wearing their new badge on operations. They knew all too well how good their chances were of being shot down and finding themselves in the company of vengeful German civilians. Bert Lee of Chelsea, Quebec, never wore the badge on his battledress after hearing about a bomb aimer who had taken to his parachute over Germany and had been executed by fear-maddened civilians upon landing. Lee had an observer's half-wing sewn on the battledress he wore on ops.) In addition to dropping bombs, the air bomber assisted the navigator, map-reading and helping to operate the *Gee, H2S,* and other newly installed equipment.

The first air bombers began training with the BCATP in August 1942. Following their ITS course, they went to bombing and gunnery schools for eight weeks. Later the course was extended to twelve weeks and the gunnery training was dropped. Their training culminated in a six-week (later ten-week) course at AOS, learning the elements of navigation. A large proportion of air bombers were washed-out pilot trainees, and many became unofficial second pilots, their skippers teaching them the basics of flying the Halifaxes and Lancasters so that they could take over in emergencies.

Another new trade for the bomber crews was that of flight engineer. Most of the first FEs were remustered RAF ground crew, men generally older than the aircrew with whom they flew. Early in 1944 the BCATP began training flight engineers in Canada. Their twenty-three week course was followed by a seven-week "type" course in Britain, where they became familiar with the aircraft in which they would fly on their squadrons. This meant that the Canadian-trained engineers would go overseas without their wings, because it was RAF policy not to bestow aircrew wings on flight engineers until they had completed their type training. Dick Reid of Montreal was one of those affected. He remembers embarking for Europe as an acting sergeant, wearing no aircrew brevet but still sporting the white flash in his cap as aircrew under training. After type training at St. Athan, Wales, the flight engineers at last received their wings – informally, in a NAAFI canteen. "Some wings parade!" comments Reid. Starting in July 1944, type training for flight engineers was carried out in Canada. The vast majority of the 1,913 flight engineers trained by the BCATP served with the squadrons of 6 (RCAF) Group on bomber operations.

The huge expansion of Bomber Command put new pressures on the BCATP to increase the output of aircrew. After demanding shorter training periods for pilots in the dark days of 1940, the British now wanted to extend the training for pilots, particularly in the area of instrument flying. The training of other members of the aircrew team also came in for reappraisal after the 1942 conference. Many of the AOSs were enlarged, with considerable increases in the numbers of students.

A few months later, in January 1943, the only naval air gunners' school in the BCATP opened at Yarmouth, Nova Scotia. No. 1 Naval Air Gunners' School (NAGS) had originated with a request from the Admiralty to Ottawa for a school to supplement those in England training telegraphist air gunners. Interestingly, the administration of the school was handled by the RCAF while the training was handled by the RN. Originally the plan was to open three training squadrons: one with forty-three Swordfish aircraft supplied by the Admiralty, the second with twenty-three Ansons from the RCAF, the third consisting of forty Stinson Reliants to be obtained in the United States. In the event, Curtiss Seamews were used at No. 1 NAGS instead of Reliants, and these aircraft in turn were replaced by Swordfish, Lysanders, Hurricanes, and a solitary Walrus also flew from No. 1 NAGS.

Robert Joss, a staff pilot at Yarmouth, recalls that "the staff pilots on the station were a mixture of RCAF officer and sergeant pilots and RN Fleet Air Arm pilots from England. The two groups got along pretty well in the wardroom which was operated as an RN officers' mess with RN traditions. Some of the RN pilots were midshipmen and there was great delight when the pilot officers in the RCAF found out that the midshipman was a commissioned officer one rank lower than a pilot officer, so life was made as miserable as possible for the Royal Navy 'snotties' in the wardroom." Joss had been posted to Yarmouth in August 1944 from the Aircrew Graduate Training School. "Needless to say," Joss says, "when the word got out that we would be flying Swordfish at Yarmouth, being hotshot fighter pilot types trained on Harvards and ready for our Hurricanes, Spitfires, etc., the thought of reverting to lumbering old biplanes with fixed undercarriages was a terrible letdown. . . . But flying any aircraft is better than not flying at all."

No. 1 NAGS graduated 704 naval air gunners before it closed in March 1945.

The BCATP had now embarked on the second phase of its existence. At the May 1942 conference, Britain agreed to provide trainees to fill "not less than 40 per cent of the training capacity of the combined training organization."[13] Australia and New Zealand (by now no longer in imminent danger of invasion) would supply a total of 4,753 trainees each year: 1,300 pilots, 676 observers, and 936 wireless operator/air gunners from Australia, and 450 pilots, 676 observers, and 715 WAG/AGs from New Zealand. Canada would supply about half the annual intake, some 34,600 trainees – although concern was expressed about the country's ability to meet this quota. The supply of aircrew reserves seemed likely to be exhausted early in 1943.

Possibly the thorniest question of the conference was that of commissions for aircrew. The Canadians declared flatly that all pilots and navigators should become officers upon graduation, with 25 per cent of other aircrew categories also being commissioned upon graduation, and another 25 per cent receiving commissions in the field. Ottawa's position was that the responsibilities undertaken by the youthful pilots and navigators of large and expensive bombers (for RAF Bomber Command was by now the principal "customer" of the BCATP) made them worthy of commissions. It was absurd, said the Canadians, for NCO pilots to command crews that included commissioned navigators, air gunners, or other aircrew; in the air, the non-commissioned pilot was in charge, yet he was supposed to address his commissioned crewmates as "sir" when on the ground. Such disparities in rank – as well as in pay and living conditions – could affect the morale of an aircrew. All the members of every crew shared the same risks. Besides, the Canadians pointed out, the promise of a commission on completion of training would surely be a fillip to recruiting.

The British saw it differently. Too many commissions would "cheapen" commissioned rank, they said. Simply because a man could fly an aeroplane or find his way about the sky didn't mean he should be an officer. Airmen received commissions because of their intelligence, character, and ability to lead and to set the right example. The existing system was flexible enough, the British felt, to provide commissions for "suitable" aircrew who had been recommended by the appropriate authorities. What irritated Canadians already on active service was the fact that "appropriate authorities" usually meant RAF officers, for the majority of Canadian airmen served in British squadrons. This fact tended to militate against the RCAF. Many RAF "brass hats" disapproved of Canadians, regarding them as rowdy cowboys who knew nothing of the subtler points of military discipline. In short, they were not the "officers and gentlemen" of legend.

The British and Canadians could not come to an agreement at the conference. So Canada adopted its own commissioning scheme. All pilots, navigators, and air bombers "considered suitable according to the standards of the Government of Canada" [14] and recommended for commissioning now became officers and enjoyed the higher rates of pay, the more attractive uniforms, the better living quarters, and the other privileges of commissioned rank.

The new system pleased the RCAF, infuriated the RAF, and created new problems for the BCATP. It meant that the training plan now had two commissioning policies: one for Canadians, and one for the British and other Commonwealth trainees. In his informative monograph on the BCATP, F. J. Hatch wrote:

> In Canada this sometimes led to invidious comparisons when, for instance, a pilot from the RAF, the RAAF, or the RNZAF, graduated at the head of his class but because of the quota was promoted to sergeant rank only, while Canadians standing further down the list received commissions. In the United Kingdom, and other theatres of war where thousands of Canadian airmen served in British squadrons, innumerable difficulties arose. In many instances the RAF commanders of these units, who were the recommending authority, were often ignorant of Canadian policy and followed the generally accepted fifty per cent rule of the RAF. Even when aware of the Canadian regulation, however, a British officer commanding a squadron in which Canadians were serving was naturally disinclined to recommend promotion to officer status of a Canadian if, in his opinion, other members of the squadron were more deserving. [15]

The Ottawa Air Training Conference ended in the early hours of June 3, 1942, its last few days brightened by the news of the mighty thousand-bomber raids carried out by Harris's aircraft. As far as the Canadians were concerned, the conference had been a notable success. They were now in total control of the BCATP, including the RAF schools, and had moved "some way down the road to maturity and independence."[16] The conference saw a new spirit of co-operation and understanding between the nations represented in the BCATP. "Chubby" Power spoke glowingly of the achievements of the conference, remarking on its "good-will," "toleration," and "friendliness and comradeship." Signed on June 5, the new BCATP agreement became effective on July 1.

Ken McDonald's tour with the BCATP ended in the summer of 1942. He had now lived in Canada for nearly three years, and had been a witness to the extraordinary progress of the BCATP. The efficiency and enthusiasm of the Canadians had impressed him enormously. He liked Canada and Canadians – and had already married one – but his tour of duty was completed. He had orders to fly a twin-engined Ventura bomber to England and join a Bomber Command squadron.

Before he left, he encountered a former RAF acquaintance, Wing Commander Spence, who had just returned from a tour with Bomber Command. McDonald told him that he was on his way to England to go on operations and asked Spence what it was like. "Mac," was the reply, "it was awful!"

Wishing he hadn't asked, McDonald set off for England, making an unscheduled stop at Eglinton, Northern Ireland, to replenish his fuel tanks. To his astonishment, the airmen at Eglinton asked him if they could have what was left over of the sandwiches he had brought for the long Atlantic flight. He soon found out why. The sandwiches had been made with white Canadian bread, not its wartime British equivalent, a curious near-beige substance, the precise ingredients of which kept an entire nation guessing for several years. The "National Loaf" was only one of the changes he found. McDonald had left England in August 1939, a couple of weeks before war broke out. His familiar world had been transformed. War and shortages had made it drab and grey. London streets bore the ugly scars of Blitz damage; the blackout became a nightly ordeal, particularly hard to take after nearly three years in the brightly lit world across the Atlantic. Prices shocked him. At the officers' mess at Pershore, he ordered a pint of beer, an ounce of St. Bruno Flake pipe tobacco, and a box of matches. He put a half-crown on the counter. In the old days it would have more than covered the purchase. Now it barely covered half of it. His England had become a place of sandbagged doorways, a land without signposts, where Anderson shelters created strange humps in tens of thousands of tiny gardens, where trees lining suburban streets looked as if they had been bandaged (the whitewashed stripes were a blackout aid for pedestrians). Air raid shelters and static water tanks dotted the landscape, and miles of familiar railings had vanished, having been chopped away to be turned into war materials. Ration books had to be produced for everything from a pork chop to a pair of flannels. Most shops had given up advertising what they had for sale; signs concentrated on telling one and all what they *didn't* have so that customers wouldn't waste their time asking and the shop assistants needn't exhaust themselves responding.

But the people hadn't changed. Although three years of war had wearied them and left their clothes shabby and worn, they brimmed with confidence. Now that the Yanks were in, the Jerries and Japs had had it, they claimed. It was just a matter of time. Radio had become of great importance in the British people's lives. They guffawed at the superstars of the medium: Tommy Handley in "ITMA" ("It's That Man Again"), and the much-loved Americans Ben Lyon and Bebe Daniels in "Hi Gang!" They danced to "Yours," "White Christmas," "Deep in the Heart of

Texas," and "I'm Gonna Get Lit Up When the Lights Go Up in London." They patronized cinemas in unprecedented numbers – and one of the most successful films of the period was *49th Parallel*, telling the story of a group of beached U-boat men trying to travel through Canada to reach the still-neutral United States.

McDonald joined 78 Squadron, a Halifax heavy bomber unit, at Linton-on-Ouse, Yorkshire, as a flight commander. He found Linton was a comfortable, prewar station, with brick buildings and central heating. Two squadrons shared the airfield: 78 and 76, the latter commanded by the remarkably durable Leonard Cheshire. McDonald describes him as a "natural leader, innovative and gregarious, with that special quality that lifts people and makes them think that in a crowd he is talking to them individually." McDonald adds that "Bomber Command was dangerous when you flew, but in between ops, Linton was an easy billet compared with most wartime stations. You had the uncomfortable feeling at the back of your mind that the next trip might be your last and I suppose that's what took the gilt off, but the daily routine wasn't that much different from peacetime. . . . Discipline was quite formal and bore the imprint of the Station Commander, John Whitley, who became an air marshal and a knight after the war and was probably one of the best-liked officers in the RAF." McDonald wrote eloquently of the impressions shared by all night bomber crews during the war:

> . . . the first sight of the stars as you broke clear of cloud after a slow, bumpy climb; the lights of Switzerland, and the Alps in moonlight; the sudden blaze of an exploding bomber; tracer weaving lazily upwards; high cloud in the distance against the moon; the sharp crack of ice particles flung from the propellers into the side of the fuselage; St. Elmo's fire ringing the propeller tips; the Halifax's buoyant leap as the bombs were released; the smell of the microphone mask and the magnified hiss of

oxygen drawn into your mouth; the brief companion-
ship of the navigator appearing beside you from his
screened compartment below; the sweep of your eyes to
left and right, down and up and across and then back to
the instruments; the red glow of another Halifax's eight
stub exhausts; the two moments of commitment, first at
take-off when you pushed all four forward and clambered into the sky, and second when someone said,
"Crossing the coast"; the forced banter in the aircrew bus
riding to dispersal before the flight and the joyous
moment afterwards when you dropped from the hatch
to Yorkshire soil again.[17]

The BCATP was about to enter its most productive period. The
huge, hastily assembled system had had time to work out its kinks.
Now it gathered momentum, turning out an increasing number
of aircrew month after month. In July 1943, the BCATP would fly
677,000 hours; in total, the year would see a record seven million
hours of flying training. Much credit for this achievement is due
to the members of the ground staff, both service and civilian. They
were the essential but largely anonymous links in the chain. Without them the BCATP's successes would have been impossible.
Their work was seldom acknowledged; much of it was monotonous, yet it could be hazardous in the extreme. Cranking recalci-
trant engines in frigid weather with ice underfoot led to several
hideous encounters with spinning propellers. Occasionally
mechanical problems created dangerous situations. At No. 33
EFTS, Caron, Saskatchewan, an RAF mechanic, Arthur Eddon,
was pulling a Tiger Moth's propeller "through" in preparation for
swinging it. Although the switches were off, the engine burst into
life. The propeller blades slashed through Eddon's coveralls,
miraculously missing the man inside. A shorting of the magnetos
was the probable cause. Such incidents occurred on every BCATP

base, and not all ground staff were as lucky as Eddon. Machine gun and bombing exercises inevitably led to mishaps, some of them fatal.

The RCAF had fewer than 1,500 fully trained tradesmen at the beginning of the war. Huge numbers had to be enlisted and trained. The Technical Training School at St. Thomas, Ontario, became the main source of ground crew, some fifty thousand in all, a large proportion of whom went to BCATP fields. In addition, the Composite Training School at Trenton provided instruction for clerks, motor transport mechanics, service police, security guards, and drill instructors; the Armament School at Mountain View, Ontario, and schools at Montreal and Clinton, Ontario, also contributed to the flow of trained ground staff. [18]

Many of the mechanics and other ground personnel at EFTSs and AOSs were civilian, about fifteen thousand in all, a third of them women. In most cases the civilian ground staff were trained by the schools, since so few aeronautical technicians were available in the early days. For example, No. 7 AOS, Portage la Prairie, began operations early in 1941 with only three ground staff possessing commercial aviation experience. The school had to train the rest, most of them former motor mechanics. At its peak, No. 7 AOS had some two hundred ground staff keeping ninety aircraft flying. [19]

By now, members of the RCAF (WD) had become familiar at BCATP fields. Formed in 1941, after air force HQ "admitted grudgingly that it was indeed feasible that many men's duties could be performed by women, even if one man's job might require two women to do it satisfactorily," [20] the Women's Division was patterned after the Women's Auxiliary Air Force (WAAF) in Britain. The WD chose as its motto the phrase, "We serve that men may fly," which today seems to reek of servility and sexism but which half a century ago apparently offended no one. The first batch of newly trained WDs went to No. 2 SFTS, Ottawa. Thereafter, WDs rapidly became a familiar sight at RCAF bases all over the country and, eventually, overseas. Some stations were still being readied for habitation when the women arrived. A WD named Beth Sinclair was posted to an eastern base. "I will never forget it, the Place of Mud! I arrived at the station in a downpour in an Air Force bus. The day was so cold and windy. When the driver stopped by the main gate to let me off, I stepped down from the bus and out of sight in a post hole full of water!"[21] Some of the first WDs thought they were going overseas when they boarded a ship at Halifax: "After three days and two nights at sea, when they had decided that they must be en route to England, they landed at St. John's and boarded a train for Gander. They learned later that they had sailed half way across the Atlantic in convoy before turning towards St. John's."[22]

Although few senior officers objected to the presence of WDs on RCAF units, many were alarmed by the added responsibility they represented. "Those who were in doubt seemed to regard us as a strange new Order which should be cloistered [for] the duration. They variously suggested the construction of high walls around our barracks, rigid partitioning in the mess halls, separate parades for airmen and airwomen, and a general programme of ascetic monasticism," wrote Wing Officer Willa Walker, one of the earliest WD graduates. "It soon became obvious that this chary attitude stemmed more from typical male uneasiness in the face of things feminine rather than from deliberate hostility."[23]

The first WDs encountered some odd reactions from airmen on the bases. One recalled: "At first the airmen were so shy that they peeked at us from behind windows and half-closed doors. But before long they were falling over one another in their efforts to make us welcome. At no time did our strength exceed a hundred which gave a most interesting ratio of about nine men to each woman."[24] Lois Argue of Regina, a medical clerk, was the first WD at No. 2 AOS, Edmonton, and she remembers being treated "rather royally" by the airmen and by the many civilians on the station. Conditions varied from unit to unit, but some WDs found the living conditions harsh and spartan. One wrote that "the barracks were bare, barren, cold and draughty. We were not allowed any kind of bedspreads, nor were we permitted to provide ourselves with curtains. Dark green blinds were at the windows. The mattresses were hard and thin."[25]

At first only eight trades were open to WDs: clerk, cook, equipment assistant, fabric worker, hospital assistant, MT driver, telephone operator, and general duties. The greatest need first appeared to be for cooks, if the alacrity with which they were enlisted is any indication. At No. 1 ANS, Rivers, Manitoba, the cooking operations were like the production line of a factory.

> Where your mother would throw a pinch of salt in the porridge at home, these people use a pailful. Their aluminum cooking pots are big enough to put two men inside. . . . To cook the bacon, a whole battery of ovens was at work. Every three minutes someone would go down the line yanking open the heavy doors and taking out the pans that were done. It would take most of the population of a piggery to provide as much bacon as was waiting on deck when the meal was ready. . . . From great stacks of cartons and boxes in the storerooms the food follows regular channels of preparation and processing until it reaches its intended destination in airmen's stomachs.[26]

The WDs were widely employed as driver/mechanics, some on ambulances. The women often had the distressing duty of driving to crash sites and helping to remove the injured and dead.

Slowly the list of trades available to WDs increased to include parachute rigger, photographer, librarian, service police, welder, wireless operator (ground), Link trainer instructor, aero-engine mechanic, and air frame mechanic. Sometimes the employment of the women caused unexpected problems. Former Warrant Officer Alexander Velleman recalls the arrival of the WDs at No. 3

B&GS, MacDonald, Manitoba, to take over the duties of wireless operators in the control tower.

> At first, we thought this was great, but we hadn't realized that problems were lying in wait. The control tower stairwells were open, that is, they had the treads on, with no risers. The new WDs insisted that the station put in risers in deference to their modesty. (At this time, the RCAF did not supply WDs with slacks, only skirts.) Further, the tower contained only two washrooms, one for men and one for male officers. With the arrival of the WDs, officers and men had to use one washroom, and the other had to be modified for use by the women. There were only three WDs at that time, and the administrative staff began to wonder if all the bother was really worth the extra help in the tower. Then the WDs came up with yet another complaint. The operators had been positioned so that they faced the runways and could fully relate to the situation. For some reason, the WDs decided that they wanted to face into the building, and thus they would require modesty panels on the operating desks. It seemed like a lot of unnecessary work, but it was done. [27]

French-speaking airwomen encountered the same problems in the unilingual RCAF as did their male confreres. The School of English at Rockcliffe, Ontario, provided a three-month course to prepare them to take training, all of which was given in English. Interestingly, no one seemed to find that fact objectionable at the time. Works minister Alphonse Fournier told the first graduates, "I am proud to see that we have in Canada young French Canadian women ready to help their country in time of need." [28] It was, he declared, an opportunity for Canada's two great races to know and understand each other better. Press reports said the course was "the best opportunity ever offered for French Canadians to become bilingual." [29] At that time in the nation's history, the idea of English-speaking Canadians learning French seemed not to occur to anyone.

"The quaint dilemma which the WDs created in the RCAF was also evidenced in the controversy over women's uniforms," wrote Willa Walker. "The question became a bone of contention both in Air Force Headquarters and among the women themselves. Although the first caps [modelled on the British WAAF model] were atrocious, we were properly womanlike and expressed our resentment when Headquarters wanted to change them. The first summer uniforms – little blue dresses with purple stockings – were so unpopular, however, that not even the contrary women objected to their discard. The men at AFHQ became very upset about these problems. In the end, the WD uniform . . . emerged second to none in both style and quality, although their evolution ironically involved the support and advice of men who ordinarily could never distinguish a Schiaparelli from a Daché." [30]

In keeping with the practices of the day, a WD received only about two-thirds of an airman's pay. An AW2 made 90 cents a day; an AC2, $1.30. Not until late in 1943 did the women's pay increase, and then it crept up to only about 80 per cent of the men's pay, hardly munificent by any standards and probably prompted only by the need to spur recruiting. "I marvel at how much we did with the measly thirty dollars a month we received," recalls Lois Argue. "We had to buy all our own lingerie and sleeping attire, all our own cosmetics . . . our entertainment, books, music, writing materials, and gifts, and we all sent big parcels overseas to the airmen every month; we donated to charity and bought War Bonds. . . ." After a brisk beginning, recruiting for the WD reached its height in July 1942 with 846 enlistments that month; thereafter recruiting declined steadily, down to 707 in October 1942 and to less than 500 a month in November and December. The following June, recruiting officers estimated that Canada had some three hundred thousand women eligible to become WDs. In an attempt to stimulate recruiting, vast numbers of posters were produced and displayed in public places. Some depicted a handsome airman in flying clothes, beaming at a pretty WD and thanking her for taking over his ground job so that he could take to the air. Others offered exciting careers in the air force with lots of travel.

Although the miserly pay undoubtedly contributed to the lacklustre recruiting record, other factors have to be considered. In general, Canadian women of that period grew up in sheltered environments, many living on farms and in small towns. The idea of signing up for the armed services and possibly going overseas to a fighting zone scared off many. Huge numbers of women remained civilians for other reasons, including the widespread rumours of immorality on RCAF bases at home and abroad. Although it is questionable whether conditions were ever as torrid as the scandalmongers made out, sexual harassment was undoubtedly a common problem for the WDs, largely because Canadians in the main preferred to believe that it didn't exist. Society had an indulgent "boys will be boys" attitude to the behaviour of servicemen in wartime, and unless outright rape was involved, complaints seldom received much consideration. Lois Argue says, "The warrant officer in the Orderly Room failed me three times on exams because I wouldn't 'party' with him – not because I didn't know the work well."

The fact was, in general, Canadians of that era tended not to approve of women being in the armed services. In 1943 a survey revealed that only about 7 per cent of the adult public thought women should be in uniform. Such attitudes made life hard for WDs. During training at No. 1 Wireless School in Montreal, Claire Fryer of Courtenay, B.C., remembers being treated rudely by the local population. On one occasion, she says, she was refused accommodation because she was in uniform. The prejudices persisted, despite the valiant efforts of recruiting officers, movies such as the National Film Board's *Proudly She Marches,* and the editorials of newspapers like the *Regina Leader-Post,* which declared ringingly that people who deliberately slandered Canada's women in uniform were not worthy to shine their shoes.

The BCATP's first operational training unit (OTU), No. 31, had become effective in August 1941 after several months of waiting for the completion of its new airfield at Debert, Nova Scotia. No. 31 was the first of seven OTUs in Canada. Back in December 1940, Lloyd Breadner, chief of the air staff, had suggested to London that OTUs should be among the units sent to Canada. Although the Air Ministry was initially cool to the suggestion, soon the benefits of such a scheme became apparent and began to change bureaucratic minds. The OTUs in Canada were already training crews for RAF Coastal Command. Why shouldn't they also prepare crews to ferry U.S.-made aircraft across the Atlantic to Britain? It is interesting to reflect that transatlantic flight was still one of the marvels of the time, so the concept of sending recent graduates in charge of expensive bombers was daring indeed. No. 31 OTU was equipped with the Lockheed Hudson, a military version of the Lockheed 14 airliner. The plan was to fly them from Dorval, near Montreal, thence to Gander, Newfoundland, from where they would set off for Prestwick, Scotland. Training began in December 1941: a twelve-week course for pilots and wireless operators, and an eight-week course for observers. The members of each crew – pilot, observer, and wireless operator/air gunner – came together in the final stage of training. They received eight more weeks of training as a crew, after which they went to Ferry Command in preparation for the Atlantic flight. The concept proved highly successful.

One of the first BCATP-trained crews to make the long trip across the Atlantic flew Hudson FH444. Bert Russell was the skipper, Jack Ritchie the navigator, and Don "Red" Macfie (John's elder brother) the wireless operator. A factory-fresh aircraft (complete with a protective sheet of paper over the navigator's table), FH444 carried no armament; the gun turret was to be installed when the aircraft arrived in Britain. Extra fuel tanks had been added in the Hudson's bomb bay. Macfie had the job of pumping the fuel from this tank to the engines: six strokes up and down per gallon. The work was hard enough in the best of conditions, but particularly exhausting if the aircraft's oxygen supply became erratic, which it did. Shortly after take-off, Russell had to climb to over twenty thousand feet to get over bad weather. Thereafter, the oxygen system knew no moderation; it delivered at full blast or not at all. The crew soon began to feel the effects of oxygen starvation. Ritchie tried to take astro shots from a hatch midway along the fuselage, since the Hudson had no astrodome. He found the task beyond him. He remembers becoming mentally sluggish and savagely irritable. At one point he came close to throwing his sextant overboard. On the other hand, Macfie found the whole episode highly amusing and couldn't stop laughing. He recalls that he "just sat and stared at the big red light on the dash, wondering what it was." Not until a fuel tank ran dry did the truth hit home. There is little doubt that any of the experienced civilian crews who had previously ferried aircraft across the Atlantic would have abandoned the flight and returned to Gander. Russell's crew never considered such a move. They had been training for this adventure for weeks and, says Jack Ritchie, regarded it as the "thrill of a lifetime."

Hudson FH444 eventually arrived at Prestwick – and, Macfie recalls, "When we landed . . . the red light was on again." It had been a near thing.

No. 36 OTU, Greenwood, Nova Scotia, was established in May 1942. The unit also operated Hudson aircraft, but it concentrated on operational training for Coastal Command; few of its graduate crews went to Ferry Command. No. 34 OTU operated Venturas, an updated but not notably successful version of the Hudson. In May 1944, No. 34 became the first of the Canadian-based OTUs to close. No. 32 OTU was based at Patricia Bay, on Vancouver Island. A torpedo-bombing unit, it trained crews for 415 Squadron RCAF (which later became a heavy bomber unit), 144 Squadron RAF, and 455 Squadron RAAF. It operated a range of aircraft – Beauforts, Hampdens, Swordfish, and Ansons – and had a brief and largely uneventful history as an operational squadron at the time of Pearl Harbor. It resumed training at the end of that eventful December in 1941. Three more OTUs appeared in the BCATP: No. 1, a fighter unit based at Bagotville, Quebec; No. 3, for the training of flying boat crews at Patricia Bay; and No. 5, a heavy bomber unit at Boundary Bay, B.C.

Although the OTUs in Canada were never a major part of the BCATP, they trained about one in ten aircrew and provided the RCAF with valuable experience for its postwar development.

BCATP graduates were now serving overseas in significant numbers. One was Vern Williams, the former accountant from The Steel Company of Canada, in Hamilton, Ontario. Now a pilot officer observer, he found himself in the pleasant English south-coast resort of Bournemouth, Hampshire, where Canadian airmen were being stockpiled until needed. For a week or two, Bournemouth could be fun; the climate was mild and the female population appealing. But the inactivity soon became tiresome. The airmen had come here to get involved in the war, not to spend their days at movies and dances. Although service wisdom had it that one should never volunteer for anything, when Williams heard that volunteers were required for a "secret project" he immediately applied, impatient to get into some sort – *any* sort – of action. He was soon on his way to an OTU at Charterhall, Scotland, where he learned to operate the still-secret AI (Airborne Interception) radar equipment then being installed on the RAF's night fighters. The early models, the Marks III and IV, had an evil reputation, constantly going u/s (unserviceable) and confronting bewildered operators with chaotic displays of dots and blobs. "It took a while to learn to interpret the mess," Williams recalls.

It was at Charterhall that Williams met the man who was to be such an important part of his subsequent air force career. Rayne Schultz had completed his flying training in Canada and had been posted overseas as a sergeant pilot. The two airmen were a study in contrasts: Williams, the quiet, teetotal accountant, and Schultz, the gregarious, fun-loving farmer's son. They joined forces and became good friends, forming a splendid night-fighting team. Soon their names would become widely known in air force circles.

Joe McCarthy from the Bronx had spent about a year with 97 Squadron, first flying Manchesters, then, without regret, moving to the excellent Lancaster. Before long he received a call from one of the leading bomber pilots in the air force, Wing Commander Guy Gibson. Gibson, who had been on operations since the outbreak of war, was in the process of forming a squadron for a special operation. Did McCarthy want to join? Without hesitation – and without knowing the purpose for which the squadron was being formed – Joe said yes. Thus he became one of Bomber Command's élite, a member of 617 Squadron, soon to be famous as the Dam Busters. Gibson had been ordered to handpick his crews for the purpose of destroying the Möhne and Eder dams with a new type of "bouncing" bomb. The job would demand superb low-level flying. The twenty-five-year-old Gibson had somehow survived 173 operational sorties at the time he formed 617. McCarthy remembers him as one of those men to whom leadership comes as naturally as breathing; autocratic and impatient at times, yet commanding instant respect. (When he visited Canada some months after the raid on the dams, Gibson made few friends in the BCATP, being critical of the pace of activity at the flying schools.) Gibson assembled a unique group of highly experienced crews for the raid, which took place on the night of May 16, 1943 – and cost almost half his squadron. Joe McCarthy came back. So did another former BCATP student, Revie Walker, who had nearly been washed out of observer training due to airsickness on one occasion. Now the navigator of Australian Dave Shannon's crew, the quiet-spoken Walker was known to everyone on the squadron as Danny. He recalls that no one knew what the target would be until a couple of days before the big day. "Most of the crews thought it was to be an attack on submarine pens," he says.

In the fall of 1943, Gibson was posted away from 617 Squadron, the unit he had led since its inception. Who could possibly fill his shoes? The squadron didn't have to wait long to find out. Joe McCarthy and Dave Shannon spotted the slim, boyish figure of their new CO near the hangars. Both pilots agreed that the new squadron commander, Leonard Cheshire by name, "didn't look like much." How wrong that first, snap judgement was to prove. McCarthy was as impressed with Cheshire as Ken McDonald had been at Linton the previous year. A superb leader, Cheshire was forever coming up with new tactics to combat the night fighters and help protect his crews. Gibson died when his Mosquito was shot down in the fall of 1944. Cheshire survived the war to found a worldwide organization for the care of the incurably ill and to become Lord Cheshire.

The air war kept costing lives, on the fighting fronts and at home. On February 25, 1943, near Fürth, Germany, a Lancaster bomber plunged into the ground, wreathed in flames. The crew died instantly, including the pilot, Don Curtin of New York City. Curtin had joined forces with Joe McCarthy to go to Canada and enlist in the RCAF. He is buried in the War Cemetery at Dürnbach, Germany. Australian Jackie Hallas, who had been surprised when a flare came in through the nose of an Anson at Hagersville, went down during a mine-laying sortie in Kiel Bay. Two weeks later his body washed ashore in Denmark. Near Armstrong, Ontario, two Ansons collided, killing the six airmen aboard. One was thirty-one-year-old Phillips Holmes, the young American actor who had given up a promising career to join the RCAF.

Pitt Clayton of Vancouver, who had joined the RAF in 1938, had had an eventful war, completing two tours of operations with Bomber Command. By mid-1943 he was a wing commander with

DFC and Bar, a wealth of experience of command – and few illusions. He had been in the RAF five years, the last two being spent in command of RCAF units: 405 and 408 Squadrons. Clayton was just the sort of combat-tested senior officer needed by the rapidly expanding RCAF, which was now the fourth largest Allied air force. Before long Clayton received an invitation to transfer from the RAF to the RCAF. He agreed. But when the authorities asked him why he wanted to switch, he raised a few high-ranking eyebrows with his candour, declaring that the reasons were twice the pay and no income tax.

The Warren twins from Nanton, Alberta, received their pilots' wings at No. 34 SFTS, Medicine Hat, Alberta. As might have been expected of these two inseparables, they graduated eighth and ninth in their class. A shock awaited them. Only the first eight received commissions. Bruce became a pilot officer, Douglas a sergeant. For the first time in their service careers, the two Dukes were separated, Bruce being posted to an RAF fighter squadron, 165. Douglas went to 403 Squadron, a Canadian unit, where his flight commander was of the opinion that brothers should not fly together. The Warrens didn't agree. Some months later Bruce persuaded his CO, Squadron Leader A. L. Winskill, that Douglas, by now commissioned, should join the squadron. To Bruce's delight, Winskill agreed. In the summer of 1942 the twins were once more together – on the ground and in the air, for they both served on the same flight, the only Canadians in the squadron. In mid-August, 165 Squadron's Spitfire Vs flew fighter cover for the bloody landings at Dieppe. Over the shell-torn beach, the twins, as part of Yellow Section, pounced on a Dornier 217 which they had spotted attacking Canadian troops. A few minutes later it plunged to destruction, engines ablaze. Only two of the four-man crew escaped. Appropriately enough the Warren twins had flown together, fought together, and had shared equally in the destruction of the enemy aircraft.

After D-Day, the twins became flight commanders in 66 Squadron RAF, as far as is known the only case of twin brothers being flight commanders in the same squadron. In the closing months of the war, the Warrens went to Buckingham Palace to receive their DFCs from King George VI. During the investiture, an aide pointed out the Warrens to the king. He looked first at one, then the other, and, as he pinned on their medals, he remarked with a smile, "I don't think I have ever done this before."

Dave Goldberg, Herb Davidson's former student, became a highly regarded instructor at No. 6 SFTS, Dunnville, then went overseas and joined 416 Squadron, flying Spitfires. In the spring of 1943 he was over France, shooting up airfields and other "targets of opportunity," when a lucky flak shot got him. He hit the ground at about 140 mph, skidding helplessly across a ploughed field until the aircraft's impetus was spent. Shaken and dazed, Goldberg wondered if he was still alive. If not, he thought, being dead wasn't all that bad. Quickly he realized that he was indeed still in the land of the living. Two civilians, farm workers by the look of them, were hurrying towards the downed Spitfire, pointing and gesticulating. There followed a pantomime in which Goldberg scrambled out of the wrecked fighter while he tried to sort out whether the French workers were pointing out the location of the Germans or indicating the direction in which he should run. He took a chance and hurried off toward some trees.

Two months later, Flying Officer Dave Goldberg arrived back in England, having escaped from occupied France via Gibraltar.

Another Spitfire pilot, former BCATP student Tet Walston, was on a PR (photo-reconnaissance) sortie at about this time. A member of 682 Squadron, Walston was flying an unarmed Spitfire XI. Not far from Genoa, Italy, a glint in the windshield-mounted rear-view mirrow caught his attention. A Messerschmitt! The

German had crept up on Walston's tail while he had been busy with his photographic work. Instinctively, he turned away. The German followed, the muzzles of his powerful cannon and machine guns all too visible in the mirror. He drew nearer . . .

Walston's mind worked at a furious pace. He remembered something. The tranquil skies of Canada . . . the instructor at Moose Jaw, the Battle of Britain veteran Flight Lieutenant Draper, DFC . . . the man with the helpful hints about what to do when pursued by a Jerry. Walston didn't pause. Slamming the throttle shut, he grasped the flap control at the top left of his instrument panel and the gear lever beside him on the right of the narrow cockpit. He jammed the first down, the second forward. Flaps and landing gear dropped, hitting the slipstream. The Spitfire felt as if it had come to an abrupt halt in the air. Wincing, ducking down in the tiny cockpit, Walston could only wait to see if the Jerry ran into him.

He didn't. He vanished into the brilliant sunlight. Walston turned for home, offering up a silent prayer for the continued good health and happiness of Flight Lieutenant Draper.

On a chilly December night, the *Luftwaffe* despatched fifty-six bombers to attack the city of Chelmsford, northeast of London. The Canadian 410 Squadron sent a solitary Mosquito II night fighter, DZ292, to investigate, the unit's other aircraft being grounded for work on the flame traps that shielded their exhaust flames at night. At DZ292's controls sat Rayne Schultz; beside him, squeezed into the narrow observer's position, was Vern Williams. They took off at 1800 hours from their base at Hunsdon, north of London, into a clear, moonlit night. While many airmen would have considered the conditions perfect, Vern Williams preferred dirty weather. "You were far less likely to be intercepted by an enemy aircraft when visibility was poor," he remarks.

Below, not a glimmer of light indicated the presence of scores of villages and towns. Schultz climbed to the patrol altitude of fifteen thousand feet. As the Mosquito headed out over the North Sea, Williams saw a flashing beacon on the Belgian-Dutch coast, which, according to a recent intelligence report, probably indicated that a raid was in progress.

Within seconds the fighter controller made contact, his voice crackling in the two airmen's earphones. "Vector 070 degrees. Investigate a bogey [unidentified aircraft] with caution." More instructions followed: "Climb to twenty thousand feet. Vector 010 degrees."

Williams's eyes were glued to his Mark V AI radar screen, on which flickering, shifting dots signified movements of aircraft in the darkness. The target was now six miles dead ahead, according to the controller. Williams picked up a contact. Schultz dived in its direction. He gathered too much speed and overshot his target. Damn! Lost him. The helpful controller gave the Canadian crew a vector of 240 degrees. Schultz turned, but the moon-streaked darkness didn't want to give up its secrets. He glimpsed the bogey for an instant, then it was gone. But not lost. Williams's radar had the enemy aircraft in its electronic grasp.

Moments later, Schultz caught sight of the German aircraft against the clouds. It was a silhouette that looked like one of those black aircraft recognition models: a Dornier 217 twin-engined bomber, shoulder-high wing, bulbous "glasshouse" in the nose, twin tails.

Schultz opened fire. The Mosquito shuddered, its tiny cabin filling with the sharp stench of cordite. One of the Dornier's engines streamed flames. It plunged, the Mosquito hard on its heels. The German pilot tried to reach cloud at seven thousand feet. He didn't make it. Another burst of gunfire sent the Dornier spiralling into the sea, still burning.

There was no time to savour the victory.

The controller's voice instructed the Canadian crew to return to fifteen thousand feet. Steer 010 degrees. Bogey three miles. Another Dornier 217.

Schultz spotted it. The distance telescoped at dizzying speed. At three hundred yards, Schultz opened fire again, the Mosquito trembling from the recoil of the cannons. With a sudden blinding roar, the Dornier blew up. Schultz flew right through the fireworks show, a dazzling instant of snapping, searing light and a monstrous eruption of power that tossed the little Mosquito about like a leaf on a stormy sea. Blackness took over once more. Aboard the Mosquito a quick check revealed that everything seemed to be in working order. No damage. Still plenty of ammo in the four 20-mm cannons. An hour of fuel in the tanks.

Williams picked up yet another contact: more "trade." It was becoming an uncommonly busy night. Starboard ten degrees. Target dead ahead. Schultz peered into the night. There was nothing to be seen, nothing but degrees of darkness, like an endless series of inky curtains opening to reveal more of the same. . . . There! At an altitude of twelve thousand feet, Schultz spotted the bogey, a third Dornier 217. He closed in for the attack.

The German pilot proved to be an airman of great skill, turning, twisting, trying every manoeuvre to evade the Mosquito on his tail and provide clear shots for his gunners.

At nine thousand feet, Schultz fired a long burst. Flames flickered about the Dornier's starboard engine. Quickly the lengthy aerial duel came down to sea level. Schultz hit the bomber again. But the enemy rear gunner was good. The Germans' return fire damaged the fighter's nose; one shot passed through the instrument panel, missing Schultz by inches. Another burst set the fighter's port engine on fire. But now both the Dornier's engines were ablaze. The bomber's crew maintained their spirited fire . . . but the game was up. The Dornier hit the sea and vanished.

The 410 Squadron Mosquito limped homeward on one engine, reaching the English coast at Bradwell Bay, touching down at 1945 hours, nearly two hours after take-off. The Mosquito's four cannons had fired 395 rounds, 125 each from three weapons, the fourth having jammed after twenty rounds. The action hadn't lasted long, probably not more than fifteen minutes from beginning to end. Yet the team of Rayne Schultz and Vern Williams had destroyed three Dorniers in a single sortie, one of the very few hat tricks scored by Allied night fighter crews in World War II. Schultz and Williams received immediate DFCs – and considerable publicity in Canada. In Hamilton, Vern Williams's mother learned of his achievement when she saw a movie newsreel in which her son was seen being decorated for his night's work.

"That's my son!" she shrieked, to the delight of the audience packed in the theatre.

CHAPTER NINE

The Instructor Factory

Fairey Battle

"When one goes to a music hall it is worthwhile thinking what a vast amount of drill the ladies of the chorus must undergo to get through their routines without a single word of command. . . . When we drill ourselves as they do there will be no more cases of pilots taking off in coarse pitch or landing without putting their wheels down."
— Robert Smith Barry[1]

John Simpson of Kingston, Ontario, graduated from the BCATP's first pilot course in the fall of 1940. He expected to be shipped overseas without delay. "When we were taking our advanced training at Borden," he recalls, "the Battle of Britain was at its peak and we all assumed that we would be in the thick of it before long."

Fate had other plans. Shortly after they received their wings, the "ugly truth," as Simpson terms it, became known to the graduates. Of the thirty-four graduates with brand-new pilots' wings on

314

their chests, seven were assigned to home defence squadrons, three went to B&GSs as staff pilots, and the twenty-four others, including Simpson, found that they were to be trained as instructors. Aptitude for the job counted for nothing, and neither did individual preferences. All that mattered was turning out as many instructors as rapidly as possible for the embryonic BCATP. "Obviously many were not suitable," Simpson says of his fellow graduates; in fact, he adds, "some were temperamentally unsuited to teach anyone anything, let alone how to fly an aeroplane." Despite widespread disappointment, the majority of those selected did their best to become capable instructors. Most succeeded. They learned their craft at Central Flying School (CFS), Trenton, Ontario, the Canadian version of the RAF's Central Flying School at Upavon, England, "the authority on flying and flying instruction for all the air forces of the Commonwealth."[2]

CFS first saw light as the Flying Instructors' School (FIS) at Camp Borden in 1939. In January 1940, the FIS moved to Trenton and a few months later acquired the name by which it became so widely known.

Simpson enjoyed the variety of aircraft he flew at CFS: "One day you would fly an Anson, a Fleet, and a Harvard. The next, a Yale, a Cessna, and a Lockheed 10 or 12." However, at that time Trenton was a grass field, and, Simpson says, "in the winter it was like landing on a frozen lake. At noon a great gaggle of aircraft would arrive, trying to get down for lunch. Ansons, Battles, Cessnas, Fleets, Harvards, and Lockheeds, all flying at different speeds, with no radios, landing four or five abreast.... You had to be alert. One morning I had a student doing an instrument take-off in a Harvard. He was slow getting the tail up and I couldn't see ahead. I gave the stick a poke . . ." The tail came up, providing Simpson with a clear view ahead. To his horror, another Harvard confronted him. It had just turned off after landing. "Because of the ice," says Simpson, "I couldn't stop or turn, so I had to go over them with the throttle through the gate. About an hour later, the same thing happened to two other Harvards, but they weren't so lucky; they collided, killing the pilot in the rear seat."

The bible at CFS was the "patter book," which detailed every flying manoeuvre on the curriculum. Instructors had to learn the patter by heart. In the air, they were required to describe manoeuvres as they demonstrated them, co-ordinating the descriptions with the actions – by no means easy without a good deal of practice. Since most of the students had just graduated after spending months learning it all, the process was in effect a matter of teaching them how to pass on what they had just learned. Simpson found it all "a bit tedious and sometimes downright irritating." One very cold January day, he was in the front seat of a Fleet, going through the sequence for steep turns. "After yelling down the voice tube for ten minutes with no reaction from the rear, I realized something was amiss. The tube was full of ice. I looked back at the student, who gave me a big smile."

The CFS became a highly efficient factory, turning out instructors by the dozen. By spring of 1942, the total exceeded 2,500. It wasn't enough. The demands of the burgeoning BCATP soon necessitated the creation of three more FISs. All three came into being in August of that year: No. 1 FIS at Trenton, a twin-engine school; No. 2 at Vulcan, Alberta, a Harvard instructors' school (later moved to Pearce, Alberta); and No. 3 at Arnprior, Ontario, near Ottawa, for elementary flying training instructors. Both Nos. 1 and 2 FIS had four testing officers and could handle sixty-six instructor-trainees at a time; the courses lasted eight weeks. At No. 3 FIS, two testing officers processed classes of twenty-six trainees in six-week courses. CFS itself became a special instructors' school and the central authority on standards of instruction in all BCATP schools.

Every student was a potential instructor. Simpson remembers always being on the lookout for students who "remained cool under adverse conditions, were really interested in what they were doing, and seemed to have the right temperament to be good instructors." Many students unwittingly determined their air force destinies by responding brightly and incisively when instructors told them to describe manoeuvres as they flew them. Instructors sometimes ordered students to "become instructors" and they would "become students" for fifteen minutes or half an hour. The students who did well could usually forget about going overseas.

Instructors were rated on their skills. The ultimate was A1, "granted only to experienced instructors of exceptional ability." [3] Next came A2, which usually went to instructors who demonstrated outstanding ability, but who perhaps lacked the experience to qualify for an A1 rating. B1 and B2 instructors might be considered the norm, excellent and thoroughly professional instructors who met all the standard requirements. A final category, C, was reserved for those who were not yet ready for the full responsibility of instructing, but who appeared to have potential. Tom Anderson of Ottawa completed an instructor's course in June 1941. He graduated and received a C rating with the following remarks: "Demonstrations given in a convincing manner – knowledge of sequences fair – should become a capable instructor with experience." Posted as an instructor to No. 2 SFTS, Ottawa, he soon had his B rating.

Most SFTS graduates had little more than two hundred hours' flying time in their logbooks when they received their wings. Those who went on to FIS to become instructors were undoubtedly capable pilots, but they lacked experience. Ideally they should have served apprenticeships, working as staff pilots; they should have been trained on high-performance aircraft as well as trainers; and they should have been given time to acquire an intimate knowledge of the air and its multitude of moods. But the war wouldn't wait. Instructors had to be created, ready or not. The wonder is not that a few of the products of the system were less than ideal but that the vast majority turned out so well. In the spring of 1943, a Trenton report criticized some instructors who had begun "to teach sequences rather than use sequences and their demonstrations to teach pupils how to fly."[4] Others, the report noted, were "artificial and unconvincing, dull and stereo-typed"[5] in their patter, failing to impress on students the connection between the lesson and the situations they might later encounter. Instructing was a daily job, often boring; inevitably, in some instructors' minds, the daily grind became the end rather than the means. On one SFTS using twin-engined aircraft, examining instructors noted that "pupils made a wild scramble in the cockpit on one engine failing, going through a form of cockpit check which obviously meant little or nothing to them."[6] At another school, instructors were criticized for not keeping up with their own instrument flying and falling into the habit of making pupils put in the required number of hours under the blind-flying hood without really teaching them to fly confidently on instruments. Similarly, some instructors taught their students how to induce spins and how to recover from them, but consistently failed to tell them about the situations that could lead to spins. Again, for a few instructors, the lessons had become more important than the *reason* for the lessons. But it should be emphasized that such instances were the exception rather than the rule in the BCATP.

About 20 per cent of trainee instructors failed to make the grade at FIS – which was precisely what most of them had been hoping for all along. "I do not recall any students who stated that they *wanted* to be instructors," John Simpson remarks. Indeed, most of those selected did their best to convince the authorities of their total unsuitability for the work. Few succeeded.

A few of these individuals took their frustrations out on their students. Alan Morris, an RAF student at No. 19 EFTS, Virden, Manitoba, had an instructor who was "skilled in his trade" but "remote, distant, unapproachable and exceedingly demanding of his trainees. . . . He always appeared to have the proverbial chip on his shoulder." Jack Baker from Lindsay, Ontario, encountered a difficult instructor at No. 7 EFTS, Windsor, who was "rude" and had "virtually no patience." He seemed to Baker to be "brassed off with the job of teaching at EFTS. . . . He swore whenever I did not follow his instructions to his liking. I still feel he deliberately set out to make me airsick . . . by doing absolutely unnecessary manoeuvres." Baker describes him as the "antithesis of what an instructor should be" – a subject Baker knew something about, for he had been a teacher in civilian life.

Arthur Bishop of New Minas, Nova Scotia, flew with an instructor at No. 6 SFTS, Dunnville, "who cursed me from the time the chocks were pulled until the propeller stopped." Frank Phripp (later a group captain in the postwar air force) remembers another instructor at Dunnville who disliked instructing and couldn't wait to get overseas. Phripp recalls: "He wasted much of my training time, which I sorely needed, in simulated dog fights with his instructor friends. I barely survived a wash-out check when a perceptive flight commander observed that I merely needed more practice." Steve Puskas of Hamilton, Ontario, did his elementary flying training at No. 9 EFTS, St. Catharines. His instructor was an American who wanted to get out of the RCAF because, according to him, he had a "job lined up with United Air Lines." Puskas recalls that his instructor consistently did his best to "be as miserable as he could be."

Some reluctant instructors went to extreme lengths to get out of instructing. John Simpson remembers a sergeant instructor at No. 3 SFTS, Calgary, who decided that it was time he went overseas. "There was a recreation area called Bowness Park which had a dance hall where we sometimes went on Saturday nights if we were not flying." recalls Simpson. "At about 2300 hours one night, an Anson appeared and really beat the place up. He was so low you could easily read the identification letters. Some civilian reported the incident, which of course was the idea. Unfortunately there had been so many fatal accidents from this sort of thing that those in charge decided to call a halt and made an example of my friend.

Instead of going overseas as he had planned, he was court-martialled, reduced to LAC rank, and posted as a staff pilot to a wireless school, flying Tiger Moths."

Fraser Gardner of St. Catharines was assigned to an instructor at No. 5 SFTS, Brantford, Ontario, who "was so negative that it is a wonder his four students ever got to wings parade." The instructor was a frustrated fighter pilot who had trained on Harvards. "The crowning insult to him," says Gardner, "was to be posted as an instructor to a twin-engined school [No. 5 SFTS operated Ansons]. He told us from day one that he would go through the motions to teach us but that his heart was not in it. Further proof of his frustration was the fact that halfway through our training he called us four students together and told us that no one was going to be recommended for a commission." The instructor enjoyed low flying from time to time. But his students derived no benefit from these expeditions, because the instructor did all the flying. "On my initiation to night cross-country flying," Gardner says, "we flew down to Detroit and then he took the controls and we went up and down Woodward Avenue [one of the city's main shopping streets] well below the tops of the buildings" – no doubt hoping to be reported by the civilian authorities.

Ironically, Fraser Gardner soon had to cope with the same frustrations as his erstwhile SFTS instructor. Posted overseas, he went to Halifax, spending three weeks at "Y" Depot, waiting to be assigned to a ship. The wait was in vain. He and about thirty others were told that they were needed as instructors – "a bitter blow," Gardner recalls. He travelled west to No. 17 SFTS, Souris, Manitoba, where the CFI ordered the newly arrived instructors to remove the "Canada" badges from their sleeves and to abandon any idea of going overseas. "He put it rather succinctly when he told us that the only way to get a posting off the station was in a box six feet long." Gardner's principal objection to instructing – one shared with innumerable colleagues – was that he was never given any indication as to how long he would be required to instruct.

Operational crews could at least look forward to the end of their tours after so many trips, but instructors – particularly NCOs – just kept going until someone told them to stop.

Toward the end of 1943, Gardner heard that nine-month tours of instruction were to be introduced. "That would have taken me to June 1944," he comments. Encouraging. But it turned out to be just another rumour. Soon Gardner heard another one: instructors would have to stay where they were, doing the same old job, because the powers that be had decided not to train any more of them. Unfortunately that one turned out to be correct. Gardner continued instructing until September 1945. He recalls that "from the fall of 1944 to the end, we were getting RAF students almost exclusively. There were a few 'repats,' chaps who had done tours as navigators or wireless operators and still wanted to get their pilots' wings."

Not all frustrated instructors were Canadians. In the fall of 1944, a number of Australian and New Zealand instructors arrived at Souris. The latter were "rather quiet and businesslike for the most part. Not so the Aussies. They became known as the 'Wild Bunch,'" Gardner says. "They did everything with gusto, including drinking beer. . . . I saw an Anson that one of them had looped. Ansons were not built for such manoeuvres. The skin of the aircraft was all wrinkled from the stress. The aircraft was a complete write-off. The Aussie pilot thought it a huge joke." But looping Ansons wasn't funny. Norman Shrive remembers an Anson crash at No. 16 SFTS, Hagersville. A high-spirited student had looped the aircraft a day or two before, cracking the main spar without realizing it. When the aircraft took to the air again, it broke up, killing an instructor and two students.

Frustration with their work led some instructors to seek solace in the bottle. Ralph Green of Grandview, Manitoba, who later became a flight commander in 424 Squadron of Bomber Command, had a civilian instructor at No. 22 EFTS, Ancienne Lorette, who drank heavily every night. "When we went flying in the morning," Green says, "his breath coming through the Gosport communications system was enough to induce impairment. Worse still, he was always too hungover to do loops, rolls, and spins, and I graduated from EFTS with whatever skills in those exercises that I was able to teach myself."

At No. 3 EFTS, London, Ontario, Nova Scotian Arthur Bishop encountered another instructor with similar tendencies, but he turned out to be a gem. Bishop's progress was above average in everything but landings. "One day in the winter of 1940/41, my instructor turned me over to another instructor in a desperate effort to get me to solo. My new instructor was a known alcoholic.... He said, 'Take off. I want to show you something.' It was a dull day with about three inches of snow on the ground. He called, 'I have control.' The next thing I knew, our wheels bounced in the middle of a farmer's field. '*There's* the goddam ground!' came the instructor's voice over the Gosport tube. Then over the hedge, and he bounced the wheels on the next field. '*There's* the goddam ground!' and on and on we went, all the way to Chatham. 'Now you have control,' he said. 'Do what I did.'"

The unusual technique proved surprisingly effective. "It must have worked," remarks Bishop, "because when we got back I 'greased one on' and he sent me solo. I had no more landing problems from that day to this."

Some instructors were cruel to be kind. Allan Caine of Toronto took twin-engine training at No. 16 SFTS, Hagersville. His instructor was "a perfectionist and always on my back, constantly criticizing me." Caine recalls that toward the end of his course the instructor was posted overseas. "One day just before he left, he said to me, 'You probably think I'm a son of a bitch. I am only trying to save your life when the going gets tough.' He was right," adds Caine, "the results of his constant demands made me a better pilot and possibly saved my life when we were forced to crash-land returning to base in England." (Caine earned the DFC serving with 420 Squadron. Later, he was saddened to learn that his former instructor had been killed overseas.)

Once a student instructor had been through the CFS course, he knew the essentials of teaching men how to fly. But the perceptive instructors, the natural teachers, knew there was more to it than that. They recognized, for example, the enormous importance of instilling confidence in their students. Joe Hartshorn instructed at No. 9 SFTS, Centralia, for some months before going overseas. To build up confidence in his students, he demonstrated his faith in their ability by apparently drinking a bottle of Coke when any of them landed the Anson aircraft. "But," he admits, "I always had my lips firmly closed outside the bottle. Just in case." Robert Brady, who had experienced engine failure shortly after his first solo at No. 17 EFTS, went on to service training at No. 8 SFTS, Moncton, switching from the docile Finch to the powerful, unforgiving Harvard. Flying Officer Pete Hamilton was his instructor. "He took note of the fact that I was a little slow in going solo in a Fleet Finch . . . so he gave me two hours in a Harvard and let me go solo – the first in the class! This gave me a tremendous confidence boost which I really never lost. He was a great guy and stuck his neck out for me." Brady and Hamilton had a close call soon after becoming acquainted. During a familiarization flight, Brady recalls, the instructor demonstrated spins and how to recover from them. "We climbed to eight thousand feet, and he put the aircraft into a spin. He calmly went through the moves, talking to me through the intercom. Down we went through six thousand feet . . . four thousand . . . three thousand . . ." At this point Brady began to notice an increasing note of concern in his instructor's voice. The earth continued to rotate, details on the ground becoming steadily larger and frighteningly distinct as the Harvard's altimeter unwound like a broken clock. Brady gulped. Time was running out . . . rapidly. Every fibre of his instructor's being was concentrated on pulling the speeding Harvard back to level flight, using emergency moves, "sawing the stick, full opposite rudder, etc., etc. . . . He was no longer calm!" The Harvard seemed destined to blow itself to bits in collision with Mother Earth, but Hamilton, using all his considerable expertise, managed to pull out of the dive at *five hundred* feet. Shaken and silent, the two men headed back to base and landed. Later, Brady and Hamilton learned to their horror that they had come close to being killed because a student pilot had been practising snap rolls and whip stalls in the Harvard, twisting its wing. Brady remarks that it took some time for him to enjoy spins again.

Instructors frequently found themselves in dangerous situations – and students weren't always involved. Bruce Brittain, from Ste. Anne de Bellevue, Quebec, became an instructor early in 1942. At Camp Borden – "still feeling like a student" – he was asked by another instructor to fly with him for some instrument work. "I asked him what he wanted to do. He replied, 'Go up to six thousand feet and fool around.'" Brittain assumed that the man meant aerobatics, which were to be recorded as instrument flying in the logbook. The other instructor did a blind take-off, his head under the blind-flying hood. "At six thousand feet, he levelled off and I released the hood. . . . We flew straight and level for a while, then started a turn; the turn got a bit tighter and we nosed down a little, picking up speed. The turn got tighter, the speed increased. I assumed he was going into high speed aerobatics. . . . At 350 mph or so, fast for a Harvard, especially if you have to pull up fast, and the ground rushing up, I started to pull out. I could feel my student-instructor start pulling as well. As we levelled out, at tree top, there was what sounded like an explosion . . . a side panel blew off . . . and both wings were rippled down to the wing roots where rivets were sprung. The question of the moment was whether the wings would fold."

The instructor thought Brittain had taken control when the latter had released the hood, although he hadn't said the customary "You have control," nor had Brittain replied with the standard "I have control." Bringing the Harvard back to earth was tricky in the extreme. Unanswerable questions kept popping into Brittain's mind. Would the starboard wing stay on? What would happen when the flaps and gear came down?

"It's normal to get some shuddering when flaps and wheels are lowered . . . in the circumstances, it wasn't too hard to imagine things falling apart . . . on top of that, were we getting proper lift with the wings buckled and the aerodynamics of the lifting surfaces upset?" Brittain had no way of knowing. He had to resist the temptation to fly the Harvard onto the ground at higher than normal speed, because of the possible stress on the airframe. He says, "No one ever concentrated more on making a smooth landing." The Harvard was a write-off, a casualty that hadn't hit anything or been hit by anything, a rare bird indeed. "What a picture it makes!" says Brittain. "Almost straight down (until we reached 350) and two pilots – *instructors* – sitting there like dummies!"

Dive-bombing and gunnery practice sometimes provided tense moments. Many students had great difficulty in the use of the gunsight, which, Brittain recalls, had "a small orange cross hair of light projected onto an angle prism, so that it was necessary to look through the prism at the target, while keeping the elusive cross hairs in sight and on the target at the same time – all the while remembering the height, which disappears very rapidly in a dive." Thus students had several important things to do simultaneously, while the ground rushed up at them at high speed. In this, as in so many aspects of flying training, instructors had to grapple with their own desire to take the controls. Instructors' instinct for self-preservation fought fierce battles with their desire not to undermine their students' self-confidence. Undoubtedly this cost some lives. Brittain had one student, a remustered WO2 mechanic who was "quite short on the kind of co-ordination required for this work. After a couple of demonstrations and several unsuccessful attempts by the student, I told him I would take a turn, but I wanted him to follow along with me – hands lightly on the controls, watching the sight but keeping his head to one side so that I could get a squint at the sight too, from the rear seat. So down we went. Unfortunately, it took a little longer to get lined up than I had bargained for, and we went far too low. It was,"

Brittain remembers, "one of those pullouts that almost leave your eyeballs hanging on your cheeks." The Harvard resumed level flight at the last instant, screaming along just a few feet above the water while instructor and student heaved heartfelt sighs of relief.

Toward the end of the summer of 1940, the CFS at Trenton instituted a quality control program for the BCATP. The concept was simple but highly effective. Groups of three or four CFS instructors called on the flying schools, staying several days to check on the progress and performance of instructors. Known as Visiting Flights, they flew with instructors – and their students – testing them on all aspects of their jobs, watching them in action, noting deficiencies, recategorizing when appropriate, usually up, occasionally down. At the end of the visit the examiners prepared detailed reports on their findings. No school took such visits lightly. Al Barton, an instructor at No. 14 SFTS, Aylmer, Ontario, recalls that the Visiting Flight pilots were "tops and were very demanding." He says, "If you were holding your own with those guys you were doing OK."[7]

The Visiting Flights did much to ensure a commendable standardization of the "products" of the BCATP flying training schools. One of the first instructors to join the Visiting Flight system was Chester Hull, whose opinion of the Cessna Crane had offended the air force hierarchy. Possessor of a coveted A1 instructor's rating, Hull became commanding officer of No. 2 Visiting Flight, at CFS, Trenton. During his tour he visited most of the BCATP's training bases, RAF as well as RCAF. In general, he recalls, the air training system operated at a high level of efficiency – astonishingly high, considering the haste and the limited resources with which the organization had been created. In the spring of 1942, Hull went to air force headquarters in Ottawa. There he joined another squadron leader and a wing commander, and this small team became responsible for training standards throughout the BCATP. At that time, he remarks, "there was no flight safety organization and it was up to us to comment on the accident reports that came in daily." (Early in 1943, Hull went overseas and did a tour of operations with 420 and 428 squadrons of Bomber Command, becoming CO of the latter unit. He attained the rank of lieutenant general in the postwar air force. His father, Allan Hull, a World War I veteran, was one of the first members of the RCAF – service number C19. He commanded No. 6 SFTS, Dunnville, during World War II.)

Scrutiny of Canadian training schools came from without as well as within. Late in 1941, Air Marshal A. G. R. Garrod, the British air member for training, inspected a number of BCATP schools. He reported that there was little to choose between the RAF and RCAF schools, describing the standard of training as "good" and the instructors as being "of high quality." School personnel, said Garrod, showed "great enthusiasm and drive in their training duties,"[8] and graduates were found to be well trained and capable, although there appeared to be room for improvement in signals, map-reading, and instrument flying. This last subject was now assuming increasing importance as more and more graduates went to Bomber Command, which operated principally at night.

The Air Ministry's high opinion of BCATP standards was short-lived. In the spring of 1943, RAF Flying Training Command began issuing quarterly reports detailing the quality of pilots trained overseas and currently taking pre-operational training in Britain. The reports came as a shock to the Canadians. The Advance Flying Units (AFUs) and OTUs in England, which prepared crews for operations, found that the ability of Canadian-trained pilots was "low in relation to the flying hours completed."[9] Other complaints included a low standard of navigation, aerobatics "generally not well performed," instrument flying "below standard," and night flying "not compatible with the hours of night flying recorded in log books."[10] Students trained in Australia, New Zealand, and South Africa generally received higher ratings from Training Command than the BCATP graduates. But it was not only in flying skills that the Canadians were found wanting. Their discipline also came in for criticism. It was a familiar theme in the saga of British-Canadian relations. Senior British officers complained that the Canadians weren't "gentlemen" – which usually meant that they didn't get into a lather of obsequiousness in the presence of lofty rank. Canadians tended to judge people by their abilities rather than by the number of stripes or pips they wore – or by their family connections. It was an attitude that had long troubled the Air Ministry, convinced as it was that gentlemen made the best aircrew, and went to the heart of the controversy between the British and Canadians on commissioning policies.

Interestingly, when RAF Coastal Command was asked to note "any marked variance . . . between those pilots trained in the UK, Canada, Southern Rhodesia, Australia," the Command responded unequivocally, reporting "no marked difference between these pilots." [11] It is perhaps worth noting that *none* of the graduate pilots arriving at Coastal Command OTUs had prior experience of the Command's OTU aircraft – Hampdens, Bothas, Beauforts, Beaufighters, and the like. Every pilot began at the same starting line. On the other hand, at Training Command bases in Britain, some Canadians may have compared poorly with Australian, New Zealand, and South African pilots because they had to fly Airspeed Oxford twin-engined trainers, a type few had seen and even fewer had flown. BCATP officers came up with another reason that their students did not always shine in comparison with those trained elsewhere: in Canada, the "cream of the EFTS crop" was often selected for single-engine training, and the best of the SFTS graduates of twin-engine schools often became instructors or were sent on general-reconnaissance courses.

Moreover, there was a puzzling lack of consistency in the quality reports. While the Australians were rated as being of a high standard in the first report, they were considered below average in the second. Clearly the quarterly reports were a far from reliable measure of the Commonwealth students. The Air Ministry decided to develop objective data.

An elaborate test took place in the fall of 1943, conducted by the Research Flight of the Empire Central Flying School. More than seven hundred pilots, all recent graduates from SFTSs in various parts of the Commonwealth, were involved: 153 from the RCAF schools in Canada, 152 from RAF schools in Canada, 105 from Southern Rhodesia, 56 from South Africa, 111 from Australia, 53 from New Zealand, 49 from British schools in the United States, and 85 who were the product of various combinations of schools – for example, Australians who had elementary flying training in Australia and service flying training in Canada. [12]

Groups of twelve pilots were tested at a time, the single-engined pilots flying Miles Master II advanced trainers – of approximately the same size and performance as the Harvard – the twin-engined pilots flying Airspeed Oxfords. Very few of the Canadian-trained pilots had experience of either of these types, but that fact seemed to have little influence on the results of the tests. Indeed, the results were remarkable only in their inconclusiveness. Certainly the tests did nothing to explain the differences between BCATP graduates and others, for the simple reason that there was so little to choose between them. The testing officers considered the standard of performance "good" and the morale of the BCATP pilots "high." [13] They gave them excellent marks for taxiing, take-off and climb, lower marks for general engine handling, aerobatics, instrument flying, and airmanship. The poorest results were achieved in the use of engine controls and instruments, overshoot procedures and control of rate of descent on instruments, and handling of ancillary controls after landing. Many Canadian-trained pilots were found to be deficient in "the checking of vital actions before and after take-off. They also showed signs of falling short of the standard expected in England in other drills and procedures." [14] On the other hand,

RCAF-trained Anson pilots were as good as RAAF Anson pilots and rather better than those trained in RAF schools. Australian Anson pilots weren't as good as the Canadians in cockpit checks, in looking out for other aircraft, in flying twin-engined aircraft on one engine, and in landing. The Australian-trained pilots were better on "vital actions after take-off" and on procedures for single-engine flying. The Australians also scored well in overshoot procedures on instruments, and their "handling of the ancillary controls after landing was decidedly better, as was their General Engine handling throughout the test." [15] Canadian Harvard pilots scored rather lower than Rhodesian Harvard pilots.

Since such an enormous amount of time and trouble went into the test, it's curious that its findings seem to have been put to such little use – if indeed *any* use was ever made of them. Perhaps the Air Ministry simply lost interest in the subject, for, by October 1944, the RAF was able to tell Ottawa that "the main deficiencies in overseas training have been eliminated, and it is becoming increasingly difficult to discriminate between the standard of training in one country as compared with another." [16] The results of the original quarterly returns, which had shaken up the senior officers of the BCATP so rudely, had long been forgotten. The *Official History* remarks: "The RAF may have preferred to transmit criticism informally and verbally. However, forty years later Air Vice-Marshal T. A. Lawrence, who held a series of important field appointments in the BCATP between 1940 and 1944 (culminating in eighteen months as air officer commanding No. 2 Training Command), could not remember any 'feedback' at all from RAF operational commands: 'There may have been some comments and criticisms passed to the RAF Liaison Officer in Ottawa, who may have passed them to RCAF Headquarters, but none were passed down to the Command level.'" [17]

An interesting experiment took place from mid-1944 to the end of January 1945 at No. 2 SFTS, Ottawa, using a group of new student pilots. The purpose was to test the practicability of training pupils in Harvards from first flight up to graduation level, without using the traditional light elementary types – Tiger Moths, Cornells, and the like. The project necessitated a special training syllabus, a combination of EFTS and SFTS curricula for a twenty-week program. Unusually bad weather extended the course to twenty-four weeks, nevertheless the overall results impressed most observers, indicating the following benefits:

1. Less flying time. Students averaged 172 hours, compared with 190 hours in the "normal" course, a saving of 9.5 per cent.
2. Less ground school time. A 20 per cent saving, from an average of 312 hours to 250 hours.
3. Less Link time. A 33 per cent saving, to 25 hours.
4. No significant difference in quality of graduates.
5. No significant difference in waste rate. The washout rate was 31.82 per cent, compared with a 30 to 35 per cent rate in combined EFTS and SFTS statistics.
6. No significant difference in the accident rate.
7. No significant difference in the cost of training. ($9,955 per "conventional" pupil versus $9,705 per "experimental" pupil.) [18]

In spite of the benefits, the idea went no further. Two of the instructors involved in the experiment were Jake Gaudaur, who had occasionally woken up some of Ottawa's citizenry with his noisy engine signals to his wife after night flying, and Peter Wrath, who obtained his A1 rating during this period. Both are of the opinion that the essential weakness of the experiment lay in the unforgiving nature of the Harvard. While a Tiger Moth or Cornell could give a student a fright if he made a mistake, a Harvard was liable to kill him. The basic trainers still had a useful role in seeing students through the uncertain early stages.

Although observers, air gunner/wireless operators, and "straight" air gunners had been trained in the BCATP since its inception, the standards had varied alarmingly at times. In fact, some aircrew were pushed through the system simply to meet quotas, conjuring up images of Soviet factory managers more interested in quantity than quality. In one instance, navigators were being graduated from general reconnaissance schools after failing their examinations, apparently a case of someone ensuring that quotas were met at all costs. In general, however, the deficiencies in BCATP standards tended to be the result of wartime shortages of equipment and personnel rather than bad management. For example, the BCATP trained thousands of air gunners with obsolete, manually operated Vickers "K" guns, when the official training manual talked glowingly of power-operated gun turrets equipped with the Browning .303 Mark II machine gun, which demonstrated such "dependability, deadly accuracy and a tremendously high rate of fire."[19] Not until late in 1942 did BCATP student air gunners have an opportunity to train on aircraft with power turrets. Even then many of their instructors lacked experience with the turrets and were unable to advise students about stoppages and other common problems.

At the beginning of the war, many senior officers still thought that the air gunner's job could be handled by anyone with a modicum of knowledge and a good eye. Even such a perceptive individual as Robert Leckie admitted to being wrong about gunners in the early days.

> When recruiting for the class started, I am well aware that I suggested that educational standards were not necessary and that the local farmer's boy who could use a gun and had the necessary guts would and should make a useful air gunner. It now appears, as a result of experience, that in this contention I was wholly wrong. . . . On planning air gunners' courses, a figure of wastage of 10%

was estimated. . . . Accumulative wastage up to October 9, 1942, was 6.5%. Since then the wastage has steadily increased from 6.5% up to 11.1%, the last figure to hand, and it is still rising. I am therefore compelled to recast courses at the B&G Schools and base them on an anticipated wastage of 20%. [20]

During the war years, Trenton acquired a forbidding reputation as a sort of halfway house for washed-out aircrew trainees. It was the place you found yourself if you didn't measure up, a place where they reasessed and redirected you, and sometimes removed you from the air force altogether. Lucien Thomas, a volunteer from Richmond, Virginia, found himself there in 1941, at the meaninglessly-named Composite Training School (known as KTS, to avoid confusion with the Conversion Training School) after failing pilot training. "I remember Trenton as the most demoralizing place I have ever been," he declares. "Failure was the byword. Everyone there was a failure in one sense or another." The downcast individuals kept arriving at Trenton, and each day brought more of them, washouts from every type and every stage of aircrew training. Still trying to accept failure, they did their best to put a brave face on things as they went about the unsavoury business of reorganizing their air force careers.

Although KTS, Trenton, seemed to most of them to be a cold, unfeeling place, a sort of recycling plant for failures, in fact a great deal of time and effort went into trying to restore the students' dented morale and put them back in the air in some capacity. For most of the war years, KTS was commanded by Wing Commander Hank Burden, a notable pilot of World War I, who shot down five enemy aircraft on one day in 1918. He became an architect between the wars. His assistant was Squadron Leader Denton Massey, former MP and socialite, who had "founded and conducted 'the world's largest' Bible class in Toronto." [21] Assisting them was a staff of interviewing officers, including former teachers, psychiatrists, and other experts. Every airman arriving at KTS received a medical re-examination, as many private interviews as seemed necessary to get to the root of his problem, and, finally, a hearing before the reselection board, which had the final say on every man's future. Three out of four of them were washed-out pilots. Most felt they should be given another chance. They had failed because their former instructors hadn't liked them. Or they had had girlfriend problems, now resolved. Or their teeth, which had ached so much during flying training, were now fixed. Or they just *knew* they wouldn't get airsick any more. Or they now saw where they had gone wrong before. They just needed another chance...

Few got what they wanted. Only about one in a hundred was sent back to resume flying training; the members of the Reselection Board knew all too well that the usual avenues had already been exhaustively explored. One of those rarities who made it back to flying training was Clyde East from Chatham, Virginia. He candidly admits that he probably made it because he was a volunteer from the United States; he doubts that the authorities would have gone to so much trouble for a Canadian in similar circumstances. During his elementary flying training at No. 4 EFTS, Windsor Mills, East was washed out, having failed a fifty-hour checkride after being caught indulging in unauthorized formation flying. East found himself at Trenton the next day. He remained there for four frustrating months, spending the time trying to persuade anyone who would listen that he had not been ready for the test and for that reason should be given another chance. To his surprise, the ploy eventually worked. East had to start flying training all over again, but this time he was careful not to break any rules along the way. The outcome was highly satisfactory for the Allied cause, for Clyde East went on to a brilliant career. He earned his pilot's wings at No. 6 SFTS, Dunnville, in November 1942. He joined 414 (Fighter Reconnaissance) Squadron RCAF, later transferring to the USAAF, shooting down thirteen enemy aircraft, and earning the remarkable total of thirty-six Air Medals during World War II and six more during the Korean War.

Some unusual cases turned up at Trenton. In the fall of 1942, a washed-out student pilot named Jim Mossman arrived. He had joined the air force wanting to become a pilot. But at his first medical, the doctor said his legs were too short for him to operate the rudder pedals of an airplane. A second doctor measured Mossman's legs in the sitting position and declared him okay for pilot training. But at No. 5 ITS, Belleville, Ontario, the doctors sided with the original MO. Mossman resigned himself to not becoming a pilot. But officialdom wasn't through with him.

"I was then called before the commanding officer of ITS, Belleville," Mossman recalls. Here, fate had a hand in the proceedings. The CO turned out to be shorter than Mossman. Understandably, he had strong views about how tall a man had to be to fly. "He stated that someone with my Link training marks was definitely going to leave his station as a pilot trainee. . . . He claimed he flew everything in the last war, and there wasn't any reason that I couldn't do the same."

The next step was Trenton, where Mossman was checked out in a Harvard. "I put on my parachute, got into the aircraft, and someone yelled, 'Full left!' and 'Full right!' I got out of the aircraft and was told, 'Legs too short.'" Back at Belleville, Mossman explained what had happened. The CO immediately informed him that he was going back to Trenton – and, he added, "This time, for God's sake, arch your back and don't let them pull you back when strapping you into your seat." Mossman passed. At last everyone was agreed: his legs were officially declared long enough for the operation of His Majesty's aircraft.

Next he went to No. 13 EFTS, St. Eugene, for pilot training. "When it came time to solo," says Mossman, "the flight commander took me out for a pre-solo check. The aircraft was a Fleet Finch. While taxiing, the flight commander yelled at me to 'gun' the motor, then yelled to 'brake' it. He then told me to return to dispersal and told me, 'Sorry, legs too short to OK a solo flight.'"

Back at Belleville, the CO told Mossman that he intended to send him to another station where Cornells were in use. He felt sure Mossman would fit its dimensions perfectly. After waiting several weeks, however, Mossman decided to apply for another aircrew trade. He became an air bomber with 429 Squadron and completed a tour of ops.

In most cases, the KTS staff set out to convince washed-out pilot trainees to remuster to some other trade – often not an easy task, for a pilot's badge was the summit of ambition for countless young men, and the notion of becoming something else was incredibly hard to swallow. Approximately 50 per cent of failed pilots became observers or air bombers. Twenty per cent more became wireless operator/air gunners, another 10 per cent "straight" air gunners. Similarly, some washed-out observers became air gunners or bomb aimers. Between 5 and 10 per cent of failed aircrew trainees reverted to ground crew or took their discharge from the air force.

Flying training became safer during the span of the BCATP, and there can be little doubt that the work of the CFS and the Accidents Investigation Branch were major factors in this achievement. Of equal importance was the simple fact that the people who ran the BCATP became increasingly proficient as the months and years went by. When the BCATP first started, many of the instructors had done little more flying than their students, and ground staff had virtually no experience of the aircraft or engines they had to maintain. In the last months, the veterans had seen it all and done it all. Perhaps the most singular fact about the statistics is that they indicate such a remarkable shift towards the end of the plan. During the fiscal years 1944 and 1945, a student was *less* likely to be killed in service flying training than he was in elementary training, an astonishing reversal of the normal state of affairs. That, however, was probably less a measure of the safety of elementary flying training than of the high spirits of young men – pupils and instructors – who couldn't resist the temptation of flying low or indulging in aerobatics that were in many cases beyond them. "Unfortunately such breaches were as often committed by flying instructors as by student pilots and the Accidents Investigation Branch reported in 1945 that 'One point was common . . . to all types of schools, namely, that trained pilots were involved in more than half of the flying accidents which occurred.'"[22]

In the first year of the BCATP, elementary flying training schools flew more than twice the hours between accidents than did service training schools: rather more than a thousand compared with just over five hundred. By the last year of the plan, the differential had been reduced to about two-thirds, and in the intervening years the two types of training had come closer to equality and had even traded places for a short period. In the period 1944-45, EFTSs were flying well over two thousand hours between accidents and SFTSs were averaging more than 1,600. The trend becomes even more pronounced when fatal accidents are considered. The first year of flying training in the BCATP saw some twenty-two thousand hours of flying between fatal accidents at EFTSs and some 8,000 hours at SFTSs. By the last two years of operation, the safety record of the EFTSs was about 10 per cent better, but the SFTSs were regularly recording well over 28,000 hours between fatal accidents, a remarkable improvement of 350 per cent.

In the BCATP, aircraft type had a considerable influence on the accident rates. Among the elementary trainers, the monoplane Fairchild Cornell (which eventually replaced most of the Tiger Moth and Fleet Finch biplanes) had an excellent accident record, despite early misgivings about its structural integrity. The twin-engined Anson and Crane, with their good-natured flying characteristics, proved themselves the safest of the advanced trainers. The more demanding Oxford, which equipped some RAF SFTSs, had a considerably higher accident rate. As might have been expected of an aircraft with its performance, the single-engined Harvard had more accidents per hours flown than any other type widely used in the BCATP.

It is interesting to note that in World War I the comparable statistics of the RFC/RAF training in Canada ranged from one fatality for every 200 hours of flying training in April 1917 to one for every 5,800 hours of training in October 1918. Again, as the experience and know-how of everyone from the commanding officer to the humblest clerk increased, so the accident rate decreased.

CHAPTER TEN

Winding Down

North American Yale

"Their shoulders held the sky suspended
They stood, and earth's foundations stay."
  – A. E. Housman. Inscription on Memorial Gates,
Trenton, honouring graduates of the BCATP who lost
their lives during World War II [1]

December 1943. The war still raged. It had become so much a part of everyday existence for so many people that it seemed to possess a life of its own. Heaven only knew when it would end. The invasion of Normandy was six months away in the uncertain future. The Germans still occupied enormous areas of Russia. Japan remained a dangerous foe in the Far East. Bomber Command was nightly losing appalling numbers of aircraft and crews in its winter campaign against Berlin and other cities. Although few in the Allied camp now doubted the eventual outcome of the war, gloomy prognosticators claimed that it could easily last another five years, after which goodness knows what sort of a world would be left.

In some ways, then, it must have seemed an odd time to start talking about terminating the BCATP. In fact it wasn't. The great buildup had been completed. By now, every month more than three thousand trained aircrew graduated and joined the dominion air forces. Pilots, navigators, gunners, and representatives of all the other aircrew trades thronged the personnel pipeline. A huge force of trained airmen had been created and, the strategists calculated, it should see the Commonwealth through to the end, with relatively few replacements needed to maintain its strength.

In February 1944, Harold Balfour, the British undersecretary of state for air, and the RAF air member for personnel, Sir Peter Drummond, arrived in Ottawa to begin the complex process of winding down the BCATP. A "ticklish business,"[2] the *Official History* described it, with all manner of implications for the Canadian economy, for the general public, and particularly for the morale of men undergoing aircrew training.

After lengthy discussions, the officials agreed that the plan should be reduced by approximately 40 per cent over the next twelve months, barring unforeseen developments in the progress of the war. March 31, 1945, was still the target date. The long-awaited invasion of occupied Europe still faced the Allies. Several people in the highest of places predicted that it would be a disaster without parallel. But most were more optimistic. And when the invasion had been successfully accomplished, the defeat of Nazi Germany would surely be just a matter of time. After that, the full might of the Allies could be directed at Japan. The RCAF planned to send twenty heavy bomber squadrons to the Far East, a force significantly larger than 6 (RCAF) Group, which represented about one-fifth the strength of RAF Bomber Command in Europe. Enormous numbers of aircrew would be needed to man this force and all the others; but the supply of trained airmen was no longer in doubt. The battle of the flying schools had been won.

The winding-down of the plan affected thousands of airmen still under training. In September 1944, Bill Mouré of Ottawa graduated as a navigator from No. 9 AOS, St. Jean, Quebec. He recalls that the members of Course 98a were promptly posted to the "Air Graduate Training School" at Calgary. In spite of its promising appellation, the school turned out to be merely a holding centre where the newly trained navigators spent their days in physical training, drill, and strenuous commando-type activities. Mouré says, "We were then informed that, because there were considerable surpluses of aircrew, it had been decided . . . 'that all aircrew now under training will be released from active service on completion of their training and transferred to the RCAF reserve.'"

Disappointed, Mouré found himself sidelined. However, the air force hadn't finished with him. In February 1945, he received a telegram ordering him to return to active service. "And so I joined the RCAF for the second time on March 15, 1945," he says. "After a quick refresher course at No. 2 ANS, Charlottetown, P.E.I., in April, I was shipped overseas, landing at Greenock on May 19, 1945, en route to the Reception Centre, Bournemouth." Mouré volunteered for service in the Pacific theatre. He was about to return to Canada for training when the Americans dropped the atomic bombs on Hiroshima and Nagasaki. The war came to an abrupt conclusion. Frustrated again, Mouré received his discharge from active service on September 14, 1945. (He joined yet again in 1948 and served six years with 435, 414, and 408 squadrons.)

Another navigator, Norman Whitley of Vancouver, had a similar experience. He graduated from the Central Navigation School, Rivers, Manitoba, in February 1945, only to be sent on leave and then discharged.

Rod Bays was an instructor at No. 10 SFTS, Dauphin, Manitoba. He recalls that toward the end of 1944 "it was becoming obvious that there were simply too many pilots in Training

Command, and there were attempts to give some sort of specialty training. There had been for some time a Standard Beam Approach Flight which attempted to teach a blind flying approach and now we were given other 'specialties', in my case the 'Bombing Flight'. . . . I think they were simply trying to make work for us. At about this time, too, new graduates were being declared 'redundant aircrew' and were sent to places like Maitland, N.S., for something known as Commando training, again making work."

The termination of the BCATP also affected a young RAF student navigator, who sailed for home without completing his training. His name had originally been Richard Jenkins, but he changed it to Burton and he hoped to get into the theatre after the war.

Most of the EFTSs had closed by December 31, 1944; only Nos. 13 and 23 remained open until the end of the war. It was a similar story with the other training establishments; one by one they shut up shop, students and staff going their separate ways, leaving cavernous hangars and empty acres behind them. Since early in the war, the bases had been a major part of so many Canadian communities. Now, abruptly, they ceased to be. No more aircraft droning overhead day and night, no more air-force-blue uniforms on the streets and in the stores, accompanied by a host of unfamiliar accents from every part of the poor old battered globe. It was all over. The end of No. 2 B&GS, Mossbank, prompted the local *Lake Johnston Star* to comment: "Well, the airport closed! This brings to all of us a feeling of melancholy. A number of the air force personnel have been with us a long time. It will be like pulling up roots to have them go." [3]

There were parades, a speech or two, hand-shaking and saluting. And here and there a few special events. Dennis Miller-Williams recalls the great toilet-roll raid on Claresholm (No. 15 SFTS): "A formation of twenty-seven Ansons gathered and set off for Claresholm on March 2, 1945. Sadly, because of a wind change and different fall rates between soft and hard paper, the missiles missed the target and drifted over the leeward boundary. However, the raid was observed and on the following day we had a low level reprisal raid by Harvards dropping flour bags."

Within a few months of the end of World War II, the BCATP was just a memory, a reminder of a bitter and brutal time that most people wanted to forget as rapidly as possible. Abandoned airfields dotted the landscape, their runways and taxiways soon cracking and spalling as nature began the long task of reasserting itself. A few bases still echoed to the clatter of aero-engines as flying clubs and schools took over. Some became postwar air force fields, undergoing all manner of modifications to ready them for the jet age. Others were transformed – utilitarian air force structures soon being replaced by dramatic new designs, architectural "statements" about the brave new postwar world and the wonders of modern air travel. Evocative names like Stevenson Field (honouring pioneer Canadian airman Fred J. Stevenson), Sea Island, and Malton vanished for ever as the former BCATP fields became international airports for Winnipeg, Vancouver, and Toronto. Other schools underwent more complete metamorphoses. A Carnation food factory stands where the hangars of No. 33 SFTS, Carberry, Manitoba, once stood. A golf course covers the acres that reverberated to the drone of engines at No. 2 B&GS, Mossbank, Saskatchewan. A detention centre for young delinquents has been constructed on the land that was once the home of No. 32 EFTS, Bowden, Alberta. A turkey farm flourishes on the former site of No. 6 SFTS, Dunnville, Ontario. In many cases, only cairns, plaques, and other memorials remain to tell anyone interested what happened there. In northern Manitoba, four lakes have been named in honour of the principal Commonwealth BCATP participants: Beaver Lake for the Canadians, Lion

Lake for the British, Kangaroo Lake for the Australians, Kiwi Lake for the New Zealanders.

To most people, the BCATP is just a dimming memory, but to those who were involved, as students, instructors, or ground staff, it is all as vivid as if it happened only a year or two ago. They sense that they were part of something unique, something that helped change history, something that will never be repeated. A measure of that sentiment must be the growing interest in reunions over the past few years. Struggling into blazers that seem to have shrunk a little since their last airing, former students and staff come from every province, every state, and from overseas to meet again and assure one another that they haven't changed a bit. The old stories emerge yet again. And every reunion brings a crop of new ones. Soon they will become part of the lore. Reg Lane and Gerry Edwards, for example, both distinguished bomber pilots during the war, became generals in the postwar air force and the best of friends when they made their homes in British Columbia after retirement. Curiously, they never discussed their training days. Then one day, years later, the subject came up. Soon they discovered that they had both done their service training at No. 10 SFTS, Dauphin, Manitoba. Lane asked which course. "Ten," was the response. Lane shook his head. Impossible. *He* was on Course 10. Now it was Edwards' turn to shake his head. No, Lane must be mistaken. Lane said he had a class photograph. A moment later the case was proved: both men were correct. In the photograph, Edwards was standing directly behind Lane; neither had been aware of the presence of the other.

While many schools have held periodic reunions, the record for consistency surely belongs to No. 6 SFTS, Dunnville, Ontario. In 1946, at an informal dinner in Galt (now Cambridge), Ontario, about a dozen former students and staff met to talk over not-so-old times. It marked the beginning of a tradition. Every year since then a No. 6 SFTS reunion has been held; enthusiasm and interest continues unabated, despite the toll exacted by the passing years.

Most of the several thousand aircraft serving the BCATP have long since been reduced to scrap. A few survive in museums such as the Commonwealth Air Training Plan Museum at Brandon, Manitoba, the Warplane Heritage Museum at Hamilton, Ontario, the RCAF Memorial Museum at Trenton, Ontario, the National Aviation Museum at Ottawa, the Air Force Museum at Comox, B.C.

By any standard, the BCATP has to be rated a spectacular achievement. Nearly half of all Commonwealth airmen received all or part of their training at BCATP schools. That the plan came into being as rapidly as it did and that it worked as efficiently as it did was little short of astounding, considering the size of the air force that had to create it and run it. "I still marvel at the organizational ability of those who made the plan a reality," comments Joffre Woolfenden, who joined the RCAF in 1938 and served for years as an instructor in the BCATP. The plan had its shortcomings, of course; the wonder is that it didn't have more. In an ideal world, every BCATP instructor would have been a combat veteran. Graduate pilots would have had at least a thousand hours of flying, much of it at night, before going overseas. All graduates would have learned the basics of other aircrew trades besides their own. All training aircraft would have been radio-equipped. All potential instructors would have been properly screened to ensure their suitability for instructing; in addition to being taught how to teach the mechanics of flying, they would have studied the psychological aspects of teaching, such as the importance of instilling self-confidence in students; they would have had tours of duty, with the promise of overseas duty after fixed periods of instructing. And the inequities of the wartime policy on commissioning still make vets' blood pressure soar.

Countless trainees criticized the BCATP for the time it took to train and process them. Bill Weighton joined the RAF in September 1942, later travelling to Canada to be trained as a navigator. By the time Weighton had graduated, travelled back to England, and had joined his operational squadron, Germany had surrendered. Similarly, Don Ripley of Orillia, Ontario, a wireless operator/air gunner, "spent two and a half years in the training/waiting mode," and the war in Europe ended just as he reached a squadron. In common with thousands of others, Ripley volunteered for service in the Pacific and was on embarkation leave when Japan surrendered.

Its shortcomings notwithstanding, the BCATP was an extraordinary achievement, one of which Canada can be immensely proud. It has rightly been described as "one of the most brilliant pieces of imaginative organization ever conceived."[4] Although more training schools operated in Britain than in Canada, there can be no doubt that the BCATP became the productive heart of the broader Empire Air Training Plan, turning out more aircrew than any of the other components.* The plan's successes completely outweighed its failures. It produced tens of thousands of aircrew when the Allies desperately needed them. And its standards kept getting higher, so that by the last months of the war, Allied airmen totally outclassed and outnumbered the enemy.

In all, the BCATP graduated 131,553 aircrew from the 159,340 who commenced training: 49,808 pilots, 29,963 navigators/observers, 15,673 air bombers, 18,496 wireless operator/air gunners, 14,996 air gunners, 1,913 flight engineers, and 704 naval air gunners.[5] The training cost the lives of more than nine hundred students, instructors, and ground staff. Two BCATP students received the George Cross, the highest non-combat award for courage. In November 1941, Tiger Moth No. 4833 from No. 2 Wireless School, Calgary, crashed in the yard of the Big Springs School, some twenty miles from base. The student WAG, Karl Gravell, was badly burned and lost one eye in the crash. Despite his injuries, he staggered back into the blazing wreckage in an

---

* The 333 aircrew schools in the Empire Air Training Plan were distributed as follows: Britain 153, Canada 92, Australia 26, South Africa 25, S. Rhodesia 10, India 9, N. Zealand 6, the Middle East 6, the United States 5, and the Bahamas 1.

attempt to free the pilot, Flying Officer James Robinson. Both men died.[6] In May 1943, Anson No. 7064 took off from No. 4 AOS, London, on a navigational exercise. The pilot was taken ill and passed out. One of the students, LAC Kenneth Spooner, immediately took charge. He ordered the other students to bail out while he handled the Anson's controls – even though he had never flown an airplane before. The Anson crashed in Lake Erie, near Port Bruce, Ontario, killing both remaining occupants.[7]

The graves of those who lost their lives with the BCATP stand as another reminder of what it all cost. At Mount Hope, Ontario, where No. 33 Air Navigation School and No. 1 Wireless School operated during the war years, local groups hold an annual service to honour the memory of the fourteen RAF airmen who died there. The tradition dates back to 1946 when the Wentworth Junior Farmers held a memorial and placed flowers on the graves. Now they are joined by veterans and air cadets. Similar ceremonies take place in many other parts of Canada.

By the latter days of the war, some BCATP graduates had become senior officers. One was Joseph P. McCarthy of Toronto (no relation to Joe McCarthy of 617 Squadron fame). Formerly a time-keeper at a plant at Malton, he had never been up in an airplane before joining the air force in July 1940. He took to flying at once. Going overseas in March 1941 as a sergeant pilot, he flew two tours of operations with Bomber Command. Promotion came rapidly to survivors in those tough days, and McCarthy rose to become the CO of 424 Squadron, a heavy bomber unit that operated in Europe and the Middle East. By the spring of 1945 he was back in Canada, a group captain. At the last wings parade held at No. 1 SFTS, Camp Borden, he pinned pilots' wings on young sergeants who experienced the greatest difficulty keeping their smiles of sheer delight under control. McCarthy knew how they felt. He had been through it.

With former students becoming senior officers and presenting wings to graduating student pilots, the BCATP, the "undertaking of great magnitude," can be said to have come full circle.

Robert Leckie, that canny, capable Scot, had planted the seed of the BCATP back in 1936 when he wrote his memorandum to Arthur Tedder about the possibility of training large numbers of Commonwealth airmen in Canada. No one is more deserving of the last word on the subject. On the tenth anniversary of the inauguration of the BCATP, he noted: "Nothing can take from us the memory of that very great adventure; although our tasks may have been unglamorous, at times monotonous, and perhaps devoid of excitement, with the passage of time we can now view our efforts in better perspective and realize that in carrying out our humble duties in Canada, we enabled others, elsewhere, to win the war." [8]

Appendix A:
The BCATP Organization

1. THE TRAINING COMMANDS

No. 1 Training Command, Toronto, Ontario
No. 2 Training Command, Winnipeg, Manitoba
No. 3 Training Command, Montreal, Quebec
No. 4 Training Command, Regina, Saskatchewan; moved to Calgary, Alberta, October 1941

2. THE ELEMENTARY FLYING TRAINING SCHOOLS

No. 1 EFTS[1]	Malton, Ont.	June 24, 1940, to July 3, 1942	Tiger Moth, Finch
No. 2 EFTS[2]	Thunder Bay, Ont.	June 24, 1940, to May 31, 1944	Tiger Moth
No. 3 EFTS[1]	London, Ont.	June 24, 1940, to July 3, 1942	Finch
No. 4 EFTS[3]	Windsor Mills, Que.	June 24, 1940, to Aug 25, 1944	Tiger Moth, Finch
No. 5 EFTS[4]	Lethbridge, Alta.; moved to High River, Alta., June '41.	July 22, 1940, to Dec 15, 1944	Tiger Moth, Cornell
No. 6 EFTS[2]	Prince Albert, Sask.	July 22, 1940, to Nov 15, 1944	Tiger Moth, Cornell
No. 7 EFTS[1]	Windsor, Ont.	July 22, 1940, to Dec 15, 1944	Finch

No. 8 EFTS[4]	Vancouver, B.C.; moved to Boundary Bay, B.C., Dec '41.	July 22, 1940, to Jan 2, 1942	Tiger Moth
No. 9 EFTS[1]	St. Catharines, Ont.	Oct 14, 1940, to Jan 14, 1944	Tiger Moth
No. 10 EFTS[1,3]	Mount Hope, Ont.; moved to Pendleton, Ont., August, 1942 (becoming part of No. 3 Command)	Oct 14, 1940, to Sept 15, 1944	Tiger Moth, Finch
No. 11 EFTS[3]	Cap de la Madeleine, Que.	Oct 14, 1940, to Feb 11, 1944	Finch, Cornell
No. 12 EFTS[1]	Goderich, Ont.	Oct 14, 1940, to Feb 11, 1944	Finch
No. 13 EFTS[3]	St. Eugene, Ont.	Oct 28, 1940, to June 19, 1945	Finch
No. 14 EFTS[2]	Portage la Prairie, Man.	Oct 28, 1940, to July 3, 1942	Tiger Moth, Finch
No. 15 EFTS[4]	Regina, Sask.	Nov 11, 1940, to Aug 11, 1944	Tiger Moth, Cornell
No. 16 EFTS[4]	Edmonton, Alta.	Nov 11, 1940, to July 17, 1942	Tiger Moth, Finch
No. 17 EFTS[3]	Stanley, N.S.	Mar 17, 1941, to Jan 14, 1944	Tiger Moth, Finch
No. 18 EFTS[4]	Boundary Bay, B.C.	April 10, 1941, to May 25, 1942	Tiger Moth
No. 19 EFTS[2]	Virden, Man.	May 16, 1941, to Dec 15, 1944	Tiger Moth, Cornell
No. 20 EFTS[1]	Oshawa, Ont.	June 21, 1941, to Dec 15, 1944	Tiger Moth
No. 21 EFTS[3]	Chatham, N.B.	July 3, 1941, to Aug 14, 1942	Finch
No. 22 EFTS[3]	Ancienne Lorette, Que.	Sept 29, 1941, to July 3, 1942	Finch
No. 23 EFTS[2]	Davidson, Sask.; moved to Yorkton, Sask., Jan '45	Nov 9, 1942, to Sept 15, 1945	Cornell

No. 24 EFTS[4]	Abbotsford, B.C.	Sept 6, 1943, to Aug 15, 1944	Cornell
No. 25 EFTS (formerly 34 EFTS)[4]	Assiniboia, Sask.	Jan 30, 1944, to July 28, 1944	Cornell
No. 26 EFTS (formerly 35 EFTS)[2]	Neepawa, Man.	Jan 30, 1944, to Aug 25, 1944	Tiger Moth
No. 31 EFTS[4]	De Winton, Alta.	June 18, 1941, to Sept 25, 1944	Stearman, Cornell
No. 32 EFTS[4]	Bowden, Alta.	July 12, 1941, to Sept 8, 1944	Stearman, Tiger Moth, Cornell
No. 33 EFTS[4]	Caron, Sask.	Jan 5, 1942, to Jan 14, 1944	Tiger Moth, Cornell
No. 34 EFTS[4]	Assiniboia, Sask.	Feb 11, 1942, to Jan 30, 1944	Tiger Moth
No. 35 EFTS[2]	Neepawa, Man.	Mar 30, 1942, to Jan 30, 1944	Tiger Moth
No. 36 EFTS[4]	Pearce, Alta.	Mar 30, 1942, to Aug 14, 1942	Stearman, Tiger Moth

[1]Part of No. 1 Training Command
[2]Part of No. 2 Training Command
[3]Part of No. 3 Training Command
[4]Part of No. 4 Training Command

3. THE SERVICE FLYING TRAINING SCHOOLS

No. 1 SFTS[1]	Camp Borden, Ont.	Nov 1, 1939, to Mar 31, 1946	Harvard, Yale
No. 2 SFTS[3]	Ottawa, Ont.	Aug 5, 1940, to April 14, 1945	Harvard, Yale
No. 3 SFTS[4]	Calgary, Alta.	Oct 28, 1940, to Sept 28, 1945	Anson, Crane
No. 4 SFTS[2]	Saskatoon, Sask.	Sept 16, 1940, to Mar 30, 1945	Anson, Crane

No. 5 SFTS[1]	Brantford, Ont.	Nov 11, 1940, to Nov 3, 1944	Anson
No. 6 SFTS[1]	Dunville, Ont.	Nov 5, 1940, to Dec 1, 1944	Yale, Harvard, Anson
No. 7 SFTS[4]	Fort Macleod, Alta.	Dec 9, 1940, to Nov 17, 1944	Anson
No. 8 SFTS[3,4]	Moncton, N.B.; moved to Weyburn, Sask., Jan '44 (becoming part of No. 4 Command)	Dec 23, 1940, to June 30, 1944	Anson, Harvard
No. 9 SFTS[3,1]	Summerside, P.E.I.; moved to Centralia, Ont., July '42 (becoming part of No. 1 Command)	Jan 6, 1941, to March 30, 1945	Anson, Harvard
No. 10 SFTS[2]	Dauphin, Man.	Mar 5, 1941, to April 14, 1945	Crane, Harvard
No. 11 SFTS[2]	Yorkton, Sask.	April 10, 1941, to Dec 1, 1944	Crane, Harvard, Anson
No. 12 SFTS[2]	Brandon, Man.	May 16, 1941, to Mar 30, 1945	Crane, Anson
No. 13 SFTS[3,2]	St. Hubert, Que.; moved to North Battleford, Sask., Feb '44 (becoming part of No. 2 Command)	Sept 1, 1941, to March 30, 1945	Anson, Harvard
No. 14 SFTS[1]	Aylmer, Ont.; moved to Kingston, Ont., August '44	July 3, 1941, to Sept 7, 1945,	Anson, Harvard, Yale
No. 15 SFTS[4]	Claresholm, Alta.	June 9, 1941, to Mar 30, 1945	Anson, Crane

No. 16 SFTS[1]	Hagersville, Ont.	Aug 8, 1941, to Mar 30, 1945	Anson, Harvard
No. 17 SFTS[2]	Souris, Man.	Mar 8, 1943, to Mar 30, 1945	Anson, Harvard
No. 18 SFTS[2]	Gimli, Man.	Sept 6, 1943, to Mar 30, 1945	Anson
No. 19 SFTS[4]	Vulcan, Alta.	May 3, 1943, to April 14, 1945	Anson
No. 31 SFTS[1]	Kingston, Ont.	Oct 7, 1940, to Aug 14, 1944	Battle, Harvard
No. 32 SFTS[4]	Moose Jaw, Sask.	Dec 9, 1940, to Oct 17, 1944	Oxford
No. 33 SFTS[2]	Carberry, Man.	Dec 26, 1940, to Nov 17, 1944	Anson, Crane
No. 34 SFTS[4]	Medicine Hat, Alta.	April 8, 1941, to Nov 17, 1944	Harvard, Oxford
No. 35 SFTS[2]	N. Battleford, Sask.	Sept 4, 1941, to Feb 25, 1944	Oxford
No. 36 SFTS[4]	Penhold, Alta.	Sept 28, 1941, to Nov 3, 1944	Oxford
No. 37 SFTS[4]	Calgary, Alta.	Oct 22, 1941, to Mar 10, 1944	Anson, Oxford, Harvard
No. 38 SFTS[4]	Estevan, Sask.	April 27, 1942, to Feb 11, 1944	Anson
No. 39 SFTS[4]	Swift Current, Sask.	Dec 15, 1941, to Mar 24, 1944	Oxford
No. 41 SFTS[4]	Weyburn, Sask.	Jan 5, 1942, to Jan 22, 1944	Anson, Harvard

[1]Part of No. 1 Training Command
[2]Part of No. 2 Training Command
[3]Part of No. 3 Training Command
[4]Part of No. 4 Training Command

4. THE AIR OBSERVER SCHOOLS

No. 1 AOS[1]	Malton, Ont.	May 27, 1940, to April 30, 1945	Anson
No. 2 AOS[4]	Edmonton, Alta.	Aug 5, 1940, to July 14, 1944	Anson
No. 3 AOS[4]	Regina, Sask.; moved to Pearce, Alta., Sept '42	Sept 16, 1940, to June 6, 1943	Anson
No. 4 AOS[1]	London, Ont.	Nov 25, 1940, to Dec 31, 1944	Anson
No. 5 AOS[1]	Winnipeg, Man.	Jan 6, 1941, to April 30, 1945	Anson
No. 6 AOS[1]	Prince Albert, Sask.	Mar 17, 1941, to Sept 11, 1942	Anson
No. 7 AOS[1]	Portage la Prairie, Man.	April 28, 1941, to Mar 31, 1945	Anson
No. 8 AOS[3]	Ancienne Lorette, Que.	Sept 29, 1941, to April 30, 1945	Anson
No. 9 AOS[3]	St. Jean, Que.	July 7, 1941, to April 30, 1945	Anson
No. 10 AOS[3]	Chatham, N.B.	July 21, 1941, to April 30, 1945	Anson

[1]Part of No. 1 Training Command
[2]Part of No. 2 Training Command
[3]Part of No. 3 Training Command
[4]Part of No. 4 Training Command

5. THE AIR NAVIGATION SCHOOLS

No. 1 ANS[1]	Trenton, Ont.; moved May 1942 to Rivers, Man., becoming Central Navigation School[2]	Feb 1, 1940, to Nov 23, 1940 Closed Sept, 1945	Anson

No. 2 ANS[2]	Pennfield Ridge, N.B.; moved to Charlottetown, P.E.I., Feb '44	July 21, 1941, to April 30, 1942 Closed July 7, 1945	Anson
No. 31 ANS[1]	Port Albert, Ont.	Nov 18, 1940, to Feb 17, 1945	Anson
No. 32 ANS[2]	Charlottetown, P.E.I.	Aug 18, 1941, to Sept 11, 1942	Anson
No. 33 ANS[1]	Mount Hope, Ont.	June 9, 1941, to Oct 6, 1944	Anson

[1]Part of No. 1 Training Command
[2]Part of No. 3 Training Command

6. THE GENERAL RECONNAISSANCE SCHOOLS

No. 1 GRS[1]	Summerside, P.E.I.	July 6, 1942, to Feb 3, 1945	Anson
No. 31 GRS[1]	Charlottetown, P.E.I.	Jan 20, 1941, to Feb 21, 1944	Anson

[1]Part of No. 3 Training Command

7. THE BOMBING AND GUNNERY SCHOOLS

No. 1 B&GS[1]	Jarvis, Ont.	Aug 19, 1940, to Feb 17, 1945	Anson, Battle, Lysander, Bolingbroke
No. 2 B&GS[4]	Mossbank, Sask.	Oct 28, 1940, to Dec 15, 1944	Anson, Battle, Lysander, Bolingbroke
No. 3 B&GS[2]	MacDonald, Man.	Mar 10, 1941, to Feb 17, 1945	Anson, Battle, Lysander, Bolingbroke
No. 4 B&GS[1]	Fingal, Ont.	Nov 25, 1940, to Feb 17, 1945	Anson, Battle, Nomad, Lysander, Bolingbroke

No. 5 B&GS[2]	Dafoe, Sask.	May 26, 1941, to Feb 17, 1945	Anson, Battle, Lysander, Bolingbroke
No. 6 B&GS[1]	Mountain View, Ont.	June 23, 1941, later becoming part of postwar RCAF	Anson, Battle, Lysander, Harvard, Nomad, Bolingbroke
No. 7 B&GS[2]	Paulson, Man.	June 23, 1941, to Feb 2, 1945	Anson, Battle, Lysander, Bolingbroke
No. 8 B&GS[4]	Lethbridge, Alta.	Oct 13, 1941, to Dec 15, 1944	Anson, Battle, Lysander, Bolingbroke
No. 9 B&GS[3]	Mont Joli, Que.	Dec 15, 1941, to April 14, 1945	Anson, Battle, Lysander, Bolingbroke, Hurricane, Norseman
No. 10 B&GS[3]	Mount Pleasant, P.E.I.	Sept 20, 1943, to June 6, 1945	Anson, Battle, Hurricane, Lysander, Norseman, Bolingbroke
No. 31 B&GS[1]	Picton, Ont.	April 28, 1941, to Nov 17, 1944	Anson, Lysander, Bolingbroke

[1]Part of No. 1 Training Command
[2]Part of No. 2 Training Command
[3]Part of No. 3 Training Command
[4]Part of No. 4 Training Command

8. THE NAVAL AIR GUNNERS' SCHOOL

No. 1 NAGS[1]	Yarmouth, N.S.	Jan 1, 1943, to Mar 30, 1945	Swordfish, Seamew, Anson, Lysander, Walrus, Hurricane

[1]Part of No. 3 Training Command

9. THE WIRELESS SCHOOLS

No. 1 WS[3,1]	Montreal, Que.; moved to Mount Hope, Ont., Sept 1944 (becoming part of No. 1 Command)	Feb 16, 1940, to Oct 31, 1945	Norseman, Tiger Moth, Stinson 105
No. 2 WS[4]	Calgary, Alta.	Sept 16, 1940, to April 14, 1945	Fort, Harvard
No. 3 WS[2]	Winnipeg, Man.	Feb 17, 1941, to Jan 20, 1945	Finch, Fort, Norseman, Yale, Tiger Moth
No. 4 WS[1]	Guelph, Ont.	July 7, 1941, to Jan 12, 1945	Tiger Moth, Norseman

[1]Part of No. 1 Training Command
[2]Part of No. 2 Training Command
[3]Part of No. 3 Training Command
[4]Part of No. 4 Training Command

10. THE FLIGHT ENGINEERS' SCHOOL

No. 1 FES[1]	Aylmer, Ont.	July 1, 1944, to Mar 31, 1945	Various aircraft used

[1]Part of No. 1 Training Command

11. THE OPERATIONAL TRAINING UNITS

No. 1 OTU[1]	Bagotville, Que.	July 20, 1942, to Jan 31, 1945	Harvard, Lysander, Hurricane, Bolingbroke
No. 3 OTU[2]	Patricia Bay, B.C.	Nov 9, 1942, to Aug 3, 1945	Stranraer, Canso, Lysander
No. 5 OTU[2]	Boundary Bay and Abbotsford, B.C.	April 1, 1944, to Oct 31, 1945	Liberator, Bolingbroke, Mitchell, Kittyhawk
No. 31 (later No. 7) OTU[1]	Debert, N.S.	June 3, 1941, to July 20, 1945	Hudson, Bolingbroke
No. 32 (later No. 6) OTU[2]	Patricia Bay, B.C., Comox, B.C., Greenwood, N.S.	June 1, 1944, later becoming part of postwar RCAF	Anson, Beaufort, Hampden, Oxford, Lysander, Dakota, Expeditor
No. 34 OTU[1]	Pennfield Ridge, N.B.	June 1, 1942, to May 19, 1944	Bolingbroke, Anson, Hudson, Ventura
No. 36 (later No. 8) OTU[1]	Greenwood, N.S.	May 11, 1942	Bolingbroke, Harvard, Hudson, Mosquito, Oxford

[1]Part of No. 3 Training Command
[2]Part of No. 4 Training Command

12. THE FLYING INSTRUCTORS' SCHOOLS

No. 1 FIS[1]	Trenton, Ont.	Aug 3, 1942, to Jan 31, 1945	Fawn, Finch, Tiger Moth, Harvard, Cornell, Crane, Anson, etc.
No. 2 FIS[3]	Vulcan, Alta.; moved to Pearce, Alta., May 1943	Aug 3, 1942, to Jan 20, 1945	Crane, Anson, Cornell, Fawn, Finch, Harvard, Oxford, Tiger Moth
No. 3 FIS[2]	Arnprior, Ont.	Aug 3, 1942, to Jan 28, 1944	Anson, Cornell, Finch, Crane, Fawn, Harvard, Stearman, Tiger Moth

[1] Part of No. 1 Training Command
[2] Part of No. 3 Training Command
[3] Part of No. 4 Training Command

13. THE SPECIALIST SCHOOLS

Central Flying School[1]	Trenton, Ont.	Feb 1, 1940, later becoming part of postwar RCAF	Anson, Battle, Bolingbroke, Fawn, Cornell, Crane, Finch, Harvard, Hudson, Hurricane, Oxford, Ventura, Lockheed 10

Central Navigation School[2]	Rivers, Man.	May 11, 1942, to Sept 15, 1945	Anson
Instrument Flying School[1]	Deseronto, Ont.	April 2, 1943, later becoming part of postwar RCAF	Oxford, Cornell

[1]Part of No. 1 Training Command
[2]Part of No. 2 Training Command

14. THE INITIAL TRAINING SCHOOLS

No. 1 ITS[1]	Toronto, Ont.
No. 2 ITS[4]	Regina, Sask.
No. 3 ITS[3]	Victoriaville, Que.
No. 4 ITS[4]	Edmonton, Alta.
No. 5 ITS[1]	Belleville, Ont.
No. 6 ITS[1]	Toronto, Ont.
No. 7 ITS[2]	Saskatoon, Sask.

[1]Part of No. 1 Training Command
[2]Part of No. 2 Training Command
[3]Part of No. 3 Training Command
[4]Part of No. 4 Training Command

15. RADIO DIRECTION FINDING (RADAR) SCHOOLS

No. 1	Leaside, Ont.
No. 31	Clinton, Ont. (renamed No. 5, July 1943)

16. OTHER SCHOOLS

Air Armament School, Mountain View, Ont.
AID Inspector School, Malton, Ont.
No. 1 Composite Training School, Trenton, Ont.
No. 2 Composite Training School, Toronto, Ont.
School of Aeronautical Engineering, Montreal, Que.
School of Aviation Medicine, Toronto, Ont.
School of Cookery, Guelph, Ont.

Appendix B:
The BCATP Aircrew Graduates

	RCAF	RAF/FLEET AIR ARM	RAAF	RNZAF	TOTAL
Pilots	25,747	17,796	4,045	2,220	49,808
Navigators	7,280	6,922	944	724	15,870
Navigators "B"	5,154	3,113	699	829	9,795
Navigators "W"	421	3,847	–	30	4,298
Air bombers	6,659	7,581	799	634	15,673
Wireless operator/air gunners	12,744	755	2,875	2,122	18,496
Air gunners	12,917	1,392	244	443	14,996
Flight engineers	1,913	–	–	–	1,913
Naval air gunners	–	704	–	–	704
Total	72,835	42,110	9,606	7,002	131,553

Note: The above covers the period from October 1940 to March 1945, but does not include 5,296 RAF and Fleet Air Arm pilots, observers, and navigators trained in Canada prior to July, 1942, principally at RAF schools, which at that period were not part of the BCATP.

Appendix C:
Flying Training Accidents

ALL TYPES OF FLYING TRAINING ACCIDENTS

Year	Hours flown per accident	
	EFTS	SFTS
1940/41	1,105	513
1941/42	751	699
1942/43	948	1,102
1943/44	1,213	1,288
1944/45	2,450	1,697

FATAL FLYING TRAINING ACCIDENTS

Year	Hours flown per accident	
	EFTS	SFTS
1940/41	22,710	8,407
1941/42	22,573	10,864
1942/43	26,917	16,782
1943/44	27,077	19,191
1944/45	25,639	28,730

ACCIDENTS BY AIRCRAFT TYPE

EFTS a/c	Hours Flown	Accident Categories*			Hrs flown per accident
		"A"	"B"	"C"	
Cornell	1,123,158	73	97	320	2,292
Tiger Moth	1,778,348	202	364	920	1,197
Finch	389,636	48	92	648	494
Stearman	75,437	11	18	53	920

SFTS a/c	Hours Flown	Accident Categories*			Hrs flown per accident
		"A"	"B"	"C"	
Anson†	4,976,431	291	505	1,108	2,614
Oxford	848,588	157	164	390	1,194
Crane	1,668,338	105	82	399	2,847
Harvard	2,968,189	343	289	2,162	1,062

*A category "A" accident was a total loss; category "B" damage was repairable by the manufacturers or at a special repair depot; category "C" damage was repairable locally.

†The figures for the Anson are from 1942-44 only, and include many thousands of hours flown by experienced staff pilots on navigated exercises.

(*Source: DHist 181.009 [D89A]*)

Notes

(n.p.n: no page numbers provided; n.p.l: no publisher listed; NAC: National Archives of Canada, Ottawa; DHist: Directorate of History, National Defence Headquarters, Ottawa.)

CHAPTER ONE: GENESIS OF THE PLAN

1. Fred H. Hitchins, *The Roundel,* December 1949, p. 13.
2. *Globe and Mail,* August 1, 1939.
3. Ibid., August 3, 1939.
4. Ron Cassels, *Ghost Squadron* (Gimli, Manitoba: Ardenlea Publishing, 1991), pp. 4-5.
5. *Globe and Mail,* August 23, 1939.
6. *Hamilton Spectator,* August 30, 1939.
7. *Globe and Mail,* August 23, 1939.
8. Brian Nolan, *King's War* (Toronto: Random House, 1988), p. 15.
9. Bruce Hutchison, *The Incredible Canadian* (Toronto: Longmans, Green & Co., 1952), p. 225.
10. *Toronto Evening Telegram,* June 30, 1937.
11. H. G. Anderson, *The Medical and Surgical Aspects of Aviation* (London: Oxford University Press, 1919), quoted in Brereton Greenhous, *A Rattle of Pebbles* (Ottawa: Dept. of National Defence, 1987), p. xiii.
12. Norman Hillmer, *Canadian Defence Quarterly,* Vol. XVI, No. 4, Spring 1987.
13. W. A. B. Douglas, *The Official History of the Royal Canadian Air Force.* Vol. II: *The Creation of a National Air Force* (Toronto: University of

Toronto Press in co-operation with the Department of National Defence and the Canadian Government Publishing Centre, Supply and Services Canada, 1986), p. 195.

14. *The Roundel,* December 1949, pp. 14-15.
15. Douglas, *Official History,* p. 195.
16. Ibid., pp. 195-6.
17. Ibid., p. 196.
18. Ibid.
19. Ibid., p. 198.
20. Ibid., p. 199.
21. Ibid., p. 202.
22. Hutchison, *Incredible Canadian,* p. 238.
23. C. P. Stacey, *A Very Double Life: The Private World of Mackenzie King* (Toronto: Macmillan of Canada, 1976), pp. 190-91.
24. Mackenzie King diaries: National Archives of Canada, Ottawa.
25. Douglas, *Official History,* p. 204.
26. Ibid., p. 205.
27. Hillmer, *Canadian Defence Quarterly,* Spring 1987, p. 52.
28. Ibid., p. 54.
29. Ibid., p. 53.
30. Douglas, *Official History,* p. 206.
31. Guy Gibson, *Enemy Coast Ahead* (London: Pan Books, 1955), p. 23.
32. F. J. Hatch, *Aerodrome of Democracy: Canada and the British Commonwealth Air Training Plan, 1939-1945* (Ottawa: Canadian Government Publishing Centre, 1983), pp. 14-15.
33. King diaries, NAC.
34. Douglas, *Official History,* p. 209.
35. King diaries, NAC.
36. Douglas, *Official History,* p. 217.
37. Hatch, *Aerodrome,* p. 17.
38. Ibid.
39. Ibid., p. 22.
40. Ibid.
41. Ibid.
42. Douglas, *Official History,* p. 216.
43. Ibid.
44. Ibid., p. 217.

45. Ibid.
46. King diaries, NAC.
47. Ibid.
48. Mackenzie King broadcast, December 17, 1939.
49. *Globe and Mail,* December 17, 1939.

CHAPTER TWO: THE GRANDIOSE ENTERPRISE

1. *The Roundel,* December 1949, p. 2.
2. Ron Cassels, *Ghost Squadron* (Gimli, Manitoba: Ardenlea Publishing, 1991), p. 1.
3. Mackenzie King diaries, NAC.
4. Leslie Roberts, *There Shall Be Wings: A History of the Royal Canadian Air Force* (Toronto: Clarke Irwin & Co., 1959), p. 100.
5. Jack C. Charleson, Canadian Aviation Historical Society *Journal,* Spring 1985, p. 5.
6. W. A. B. Douglas, *The Official History of the Royal Canadian Air Force.* Vol. II: *The Creation of a National Air Force* (Toronto: University of Toronto Press in co-operation with the Department of National Defence and the Canadian Government Publishing Centre, Supply and Services Canada, 1986), p. 141.
7. Directorate of History, National Defence Headquarters, Ottawa, file No. 80/395.
8. Ibid.
9. Ibid.
10. Ibid.
11. DHist. 74/20.
12. DHist. 181.003(D4776).
13. E. C. Luke, *The Roundel,* December 1949, p. 16.
14. Ibid.
15. DHist. 74/20.
16. Luke, *The Roundel,* p. 17.
17. DHist. 181.003(D4776).
18. F. J. Hatch, *Aerodrome of Democracy: Canada and the British Commonwealth Air Training Plan, 1939-1945* (Ottawa: Canadian Government Publishing Centre, 1983), p. 36.

19. R. V. Manning, *The Roundel,* July/August 1960, pp. 16-17.
20. Douglas, *Official History,* p. 231.
21. E. C. Luke, *The Roundel,* December 1949, p. 16.
22. *Fortune,* April 1942, p. 84.
23. K. M. Molson and H. H. Taylor, *Canadian Aircraft Since 1909* (Stittsville, Ontario: Canada's Wings, 1982), p. 58.
24. Leslie Roberts, *Canada's War in the Air* (Montreal: Alvah M. Beatty Publications, 1943), p. 47.
25. Alex McAlister, *Sky-High!* (Toronto: The Ryerson Press, 1944), p. 7.
26. Dave McIntosh, *Terror in the Starboard Seat* (Don Mills: General Publishing, 1980), p. 13.
27. Norman Shrive, "The Way Some of Us Were," Part III, *Flightlines,* published by the Canadian Warplane Heritage Museum, April 1991, p. 64.
28. DHist. 181.009(D89A).
29. Murray Peden, *A Thousand Shall Fall* (Toronto: Stoddart Publishing Co., 1988), p. 16.
30. Shrive, "The Way," Part IV , *Flightlines,* Fall 1991, p. 31.
31. McAlister, *Sky-High!,* p. 26.
32. Peden, *A Thousand,* pp. 27-28.
33. McAlister, *Sky-High!,* p. 29.

CHAPTER THREE: PILOT TRAINING: THE BEGINNING

1. Norman Shrive, "The Way Some of Us Were," Part V, *Flightlines,* published by the Canadian Warplane Heritage Museum, Spring 1992, p. 20.
2. William A. Martin, "Letters to a Kiwi," an unpublished memoir, p. 27.
3. F. J. Hatch, *Aerodrome of Democracy: Canada and the British Commonwealth Air Training Plan, 1919-1945* (Ottawa: Canadian Government Publishing Centre, 1983), p. 117.
4. DHist. 181.009(D89A).
5. Ibid.
6. Len Morgan, *The AT-6 Harvard* (New York: Arco Publishing Co., 1965), n.p.n.
7. Alex McAlister, *Sky-High!* (Toronto: The Ryerson Press, 1944), p. 41.

8. F. D. Tredrey, *Pioneer Pilot* (London: Peter Davies, 1976), p. 16.
9. Sheila Hailey, *I Married a Best Seller* (New York: Doubleday & Co., 1978), p. 29.
10. John Macfie diary, NAC.
11. DHist. 181.009(D6502).
12. DHist. 181.009(D89A).
13. Ibid.
14. Ibid.
15. Morgan, *Harvard*, n.p.n.
16. Peden, *A Thousand Shall Fall* (Toronto: Stoddart Publishing Co., 1988), p. 34.
17. Morgan, *Harvard*, n.p.n.
18. Ibid.
19. McAlister, *Sky-High!*, pp. 51-2.
20. Martin, "Kiwi," p. 27.
21. Robert Brady, "Great Little Biplane," *Airforce*, September 1992, p. 42.
22. Martin, "Kiwi," p. 32.
23. McAlister, *Sky-High!*, p. 58.

CHAPTER FOUR: ADVANCED FLYING TRAINING

1. F. D. Tredrey, *Pioneer Pilot* (London: Peter Davies, 1976), p. 98.
2. Len Morgan, *The AT-6 Harvard* (New York: Arco Publishing Co., 1965), n.p.n.
3. Jack Merryfield, *Contact*, Commonwealth Air Training Plan Museum Inc. newsletter, Brandon, Man., January 1987, p. 3.
4. DHist. 181.009(D4761).
5. DHist. 80/408.
6. Morgan, *Harvard*, n.p.n.
7. Alex McAlister, *Sky-High!* (Toronto: The Ryerson Press, 1944), p. 94.
8. Ibid., p. 96.
9. John Macfie diary, NAC.
10. Morgan, *Harvard*, n.p.n.
11. Lew Duddridge, *The Best Seventy Years of My Life* (Victoria: Orca Book Publishers, 1988), p. 56.
12. Max Ward, *The Max Ward Story* (Toronto: McClelland & Stewart, 1991), pp. 31-2.

13. Peter Townsend, *Duel of Eagles* (New York: Simon and Schuster, 1970), p. 252.
14. Keith Park, *Battle of Britain* (quoted in *Glorious Summer* by Johnson and Lucas, St. Paul, 1990, p. 106).
15. Richard Rivaz, *Tail Gunner Takes Over* (London: Jarrolds, 1945), pp. 83-5.
16. Arthur Bishop, *The Courage of the Early Morning* (London: William Heinemann, 1966), p. 187.
17. Bert Houle video interview: RCAF Memorial Museum, Trenton.

CHAPTER FIVE: THE OTHER AIRCREW TRADES

1. F. J. Hatch, *Aerodrome of Democracy: Canada and the British Commonwealth Air Training Plan, 1939-1945* (Ottawa: Canadian Government Publishing Centre, 1983), p. 163.
2. Larry Milberry and Hugh A. Halliday, *The Royal Canadian Air Force at War, 1939-1945* (Toronto: Canav Books, 1990), p. 67.
3. Ibid.
4. DHist. 73/1558 Vol. 1.
5. Ibid.
6. Ibid.
7. Ibid.
8. Ibid.
9. Frank Covert, unpublished memoirs, courtesy of W. C. Pierce.
10. Ron Cassels, *Ghost Squadron* (Gimli, Manitoba: Ardenlea Publishing, 1991), p. 15.
11. Don Charlwood, *Journeys Into Night* (Victoria, Australia: Hudson Publishing, 1991), pp. 24-5.
12. Cassels, *Ghost Squadron,* pp. 15-16.
13. Chaz Bowyer, *Guns in the Sky* (London: Corgi Books, 1981), pp. 35-6.
14. *Fortune,* April 1942, p. 146.
15. W. A. B. Douglas, *The Official History of the Royal Canadian Air Force.* Vol. II: *The Creation of a National Air Force* (Toronto: University of Toronto Press in co-operation with the Department of National Defence and the Canadian Government Publishing Centre, Supply and Services Canada, 1986), p. 237.

CHAPTER SIX: TOWN AND COUNTRY

1. Brereton Greenhous and Norman Hillmer, "The Impact of the BCATP on Western Canada: Some Saskatchewan Case Studies," *Journal of Canadian Studies*, Vol. 16, Nos. 3 and 4, Fall/Winter 1981, p. 134.
2. Ibid.
3. Ibid.
4. *Trenton Courier-Advocate*, October 12, 1929.
5. *Journal of Canadian Studies*, Vol. 16, p. 134.
6. Pierre Berton, *The Great Depression, 1929-1939* (Toronto: McClelland & Stewart, 1990), p. 247.
7. *Journal of Canadian Studies*, Vol. 16, p. 135.
8. Ibid., p. 135.
9. Ibid., pp. 135-6.
10. *Furrows and Faith: A History of Lake Johnston and Sutton Rural Municipalities*, (n.p.l.), 1980, p. 371.
11. *Journal of Canadian Studies*, Vol. 16, p. 136.
12. Ibid., p. 136.
13. Ibid.
14. Eugene McGee, *History of Port Albert* (n.p.l.), p. 14.
15. *Message to Base* (Newsletter: April 1945), p. 7.
16. Alexander Velleman, *The RCAF as Seen From the Ground (A Worm's-Eye View)*. Book I: *The War Years, 1939-1945* (Stittsville, Ontario: Canada's Wings, 1986), p. 137.
17. *Dunnville Chronicle*, September 23, 1992.
18. *Journal of Canadian Studies*, Vol. 16, p. 137.
19. Ibid., p. 137.
20. Roger A. Freeman, *The British Airman* (London: Arms and Armour Press, 1989), pp. 19-20.
21. *Winnipeg Free Press*, June 3, 1943.
22. *Journal of Canadian Studies*, Vol. 16, p. 137.
23. William A. Martin, "Letters to a Kiwi," an unpublished memoir, p. 32.
24. *Journal of Canadian Studies*, Vol. 16, p. 141.
25. Ibid.
26. *Regina Leader-Post*, September 13, 1944.

27. *Prairie Flyer* (Newsletter: various issues).
28. *Regina Leader-Post,* September 13, 1944.
29. F. J. Hatch, *Aerodrome of Democracy: Canada and the British Commonwealth Air Training Plan, 1939-1945* (Ottawa: Canadian Government Publishing Centre, 1983), p. 55.
30. Don Charlwood, *Journeys Into Night* (Victoria, Australia: Hudson Publishing, 1991), p. 16.

CHAPTER SEVEN: OVER HERE

1. Franklin Delano Roosevelt, "Fireside Chat" radio broadcast, December 29, 1940.
2. Gregory Ross, *America 1941* (New York: The Free Press, 1989), p. 17.
3. Ibid., p. 20.
4. Ibid., p. 21.
5. Ibid., p. 22.
6. *Newsweek,* March 16, 1940.
7. *Saturday Evening Post,* February 1, 1941.
8. Ibid.
9. Ibid.
10. *Collier's,* April 20, 1940.
11. Quoted in Stuart Leuthner and Oliver Jensen, *High Honor: Recollections by Men and Women of World War II Aviation* (Washington: Smithsonian Institute, 1989), pp. 324-5.
12. U.S. Department of State, letter to the author, November 26, 1993, ref. PA/HO.
13. DHist. 80-68 File 2.
14. Ibid.
15. Arthur Bishop, *The Courage of the Early Morning* (London: William Heinemann, 1966), p. 184.
16. DHist. 80-68 File 2.
17. Hatch, *Aerodrome,* p. 87.
18. DHist. 80-68 File 2.
19. Ibid.
20. Ibid.
21. Ibid.

22. Ibid.
23. Ibid.
24. Ibid.
25. Ibid.
26. DHist. 80-60. Interview with Clayton Knight.
27. DHist. 80-68 File 2.
28. DHist. 80-68 File 2.
29. Ibid.
30. Ibid.
31. Ibid.
32. Canadian National Geographic *Journal,* December 1941, p. 299.
33. Ibid., p. 283.
34. Information re J. W. G. Clark courtesy of Joe Clark, Jr.
35. Massey to Warner, February 5, 1941, Warner Brothers archives, University of Southern California.
36. Dorothy Massey to Wallis, February 5, 1941, Warner archives, USC.
37. Rudy Mauro, "The Making of Captains of the Clouds," Canadian Aviation Historical Society *Journal,* summer 1991, p. 42.
38. Power to Raine, April 1, 1941, Warner archives, USC.
39. Wald to Wallis, April 17, 1941, Warner archives, USC.
40. Draft script, Warner archives, USC.
41. Ibid.
42. Production notes, Warner archives, USC.
43. Ibid.
44. Ibid.
45. Ibid.
46. Bishop, *Courage of the Early Morning,* p. 189.
47. Interview with Arthur Bishop, December 17, 1992.
48. James Cagney, *Cagney by Cagney* (New York: Doubleday & Co., 1976), p. 102.
49. Production note, Warner archives, USC.
50. Ibid.
51. Stacey to Hudson's Bay Co., North Bay, July 15, 1941, Warner archives, USC.
52. H. Hugh Wynne, *The Motion Picture Stunt Pilots* (Missoula, Montana: Pictorial Histories Publishing Co., 1987), p. 163.

53. *Variety Film Reviews,* January 21, 1942.

54. Quoted by Rudy Mauro, Canadian Aviation Historical Society *Journal,*
    Fall 1991, p. 95.

55. Ibid.

CHAPTER EIGHT: THE ROAD TO MATURITY

1. Donald Creighton, *The Forked Road: Canada 1939-1957* (Toronto:
   McClelland & Stewart, 1976), p. 89.

2. Ibid., p. 46.

3. W. A. B. Douglas, *The Official History of the Royal Canadian Air Force.*
   Vol. II: *The Creation of a National Air Force* (Toronto: University of
   Toronto Press in co-operation with the Department of National
   Defence and the Canadian Government Publishing Centre, Supply
   and Services Canada, 1986), p. 251.

4. Ibid.

5. Ibid.

6. Ibid., p. 252.

7. Mackenzie King diaries, NAC.

8. Douglas, *Official History,* pp. 257-8.

9. Ibid., p. 258.

10. Ibid., p. 259.

11. Ibid., p. 260.

12. Ibid.

13. F. J. Hatch, *Aerodrome of Democracy: Canada and the British
    Commonwealth Air Training Plan, 1939-1945* (Ottawa: Canadian
    Government Publishing Centre, 1983), p. 106.

14. Ibid., p. 108.

15. Ibid.

16. Douglas, *Official History,* p. 263.

17. Ken McDonald, "Recollections of Linton-on-Ouse," *Airforce,*
    September 1981, p. 16.

18. Skelding and Elliott, *The Roundel,* December 1949, p. 19.

19. *The Record* (Portage la Prairie: Portage Air Observer School Ltd.,
    1945), pp. 42-3.

20. *We Serve That Men May Fly* (Hamilton: RCAF [WD] Association,
    1973), p. 6.

21. Ibid., p. 34.
22. Ibid., p. 36.
23. Willa Walker, *The Roundel*, December 1949, p. 38.
24. *We Serve*, p. 36.
25. Ibid., p. 37.
26. Ibid., pp. 38-9.
27. Alexander Velleman, *The RCAF As Seen From the Ground (A Worm's-Eye View)*. Book I: *The War Years, 1939-1945* (Stittsville, Ontario: Canada's Wings, 1986), pp. 74-5.
28. *We Serve*, p. 55.
29. Ibid.
30. Willa Walker, *The Roundel*, December 1949, p. 39.

CHAPTER NINE: THE INSTRUCTOR FACTORY

1. Quoted in Tredrey, *Pioneer Pilot* (London: Peter Davies, 1976), p. 126.
2. F.J. Hatch, *Aerodrome of Democracy: Canada and the British Commonwealth Air Training Plan, 1939-1945* (Ottawa: Canadian Government Publishing Centre, 1983), p. 116.
3. Ibid.
4. W. A. B. Douglas, *The Official History of the Royal Canadian Air Force*. Vol. II: *The Creation of a National Air Force* (Toronto: University of Toronto Press in co-operation with the Department of National Defence and the Canadian Government Publishing Centre, Supply and Services Canada, 1986), p. 268.
5. Ibid.
6. Ibid. p. 269.
7. Larry Milberry and Hugh A. Halliday, *The Royal Canadian Air Force at War, 1939-1945* (Toronto: Canav Books, 1990), p. 48.
8. Douglas, *Official History*, p. 269.
9. Ibid., p. 270.
10. Ibid.
11. Ibid., p. 271.
12. Ibid., p. 272.
13. Ibid., p. 273.
14. Ibid.

15. Ibid.
16. Ibid., p. 274.
17. Ibid.
18. DHist. 181.009(D89A).
19. *The Service Aircrew,* service manual, 2nd ed., 1943, Armament, p. 12.
20. Douglas, *Official History,* p. 276.
21. *Fortune,* April 1942, p. 148.
22. DHist. 181.009(D89A).

CHAPTER TEN: WINDING DOWN

1. *The Roundel,* December 1949, p. 2.
2. W. A. B. Douglas, *The Official History of the Royal Canadian Air Force.* Vol. II : *The Creation of a National Air Force* (Toronto: University of Toronto Press in co-operation with the Department of National Defence and the Canadian Government Publishing Centre, Supply and Services Canada, 1986), p. 292.
3. *Lake Johnston Star,* December 1944, quoted in *Furrows and Faith: A History of Lake Johnston and Sutton Rural Municipalities,* (n.p.l.), 1980, p. 371.
4. Maurice Dean, *The RAF and Two World Wars* (London: Cassell, 1979), p. 77.
5. F. J. Hatch, *Aerodrome of Democracy: Canada and the British Commonwealth Air Training Plan, 1939-1945* (Ottawa: Canadian Government Publishing Centre, 1983), p. 206.
6. Les Allison and Harry Hayward, *They Shall Not Grow Old* (Brandon: Commonwealth Air Training Plan Museum, 1992), p. 278.
7. Ibid., pp. 719-20.
8. *The Roundel,* December 1949, p. 3.

Bibliography

Allison, Les, and Harry Hayward. *They Shall Not Grow Old.* Brandon: Commonwealth Air Training Plan Museum, 1992.

Berton, Pierre. *The Great Depression, 1929-1939.* Toronto: McClelland & Stewart, 1990.

Bishop, Arthur. *The Courage of the Early Morning.* London: William Heinemann, 1966.

Bowyer, Chaz. *History of the RAF.* London: Hamlyn, 1977.

———. *Guns in the Sky.* London: Corgi Books, 1981.

Cagney, James. *Cagney by Cagney.* New York: Doubleday & Co., 1976.

Cassels, Ron. *Ghost Squadron.* Gimli, Manitoba: Ardenlea Publishing, 1991.

Charlwood, Don. *Journeys Into Night.* Victoria, Australia: Hudson Publishing, 1991.

Creighton, Donald. *The Forked Road: Canada 1939-1957.* Toronto: McClelland & Stewart, 1976.

Douglas, W. A. B. *The Official History of the Royal Canadian Air Force.* Vol. II: *The Creation of a National Air Force.* Toronto: University of Toronto Press in co-operation with the Department of National Defence and the Canadian Government Publishing Centre, Supply and Services Canada, 1986.

Duddridge, Lew. *The Best Seventy Years of My Life.* Victoria, B.C.: Orca Book Publishers, 1988.

Fletcher, David C., and Doug MacPhail. *Harvard! The North American Trainers in Canada.* Dundas, Ont.: DCF Flying Books, 1990.

Freeman, Roger A. *The British Airman.* London: Arms and Armour Press, 1989.

Gibson, Guy. *Enemy Coast Ahead.* London: Pan Books, 1970.

Greavette, Major G. E. I. *Fifty Years of Flying Training.* Portage la Prairie: Manitoba Airshow, 1990.

Greenhous, Brereton. *A Rattle of Pebbles: The First World War Diaries of Two Canadian Airmen.* Ottawa: Canadian Government Publishing Centre, 1987.

Gregory, Ross. *America 1941.* New York: The Free Press, 1989.

Hailey, Sheila. *I Married a Best Seller.* New York: Doubleday & Company, 1978.

Hatch, F. J. *Aerodrome of Democracy: Canada and the British Commonwealth Air Training Plan, 1939-1945.* Ottawa: Canadian Government Publishing Centre, 1983.

Hutchison, Bruce. *The Incredible Canadian.* Toronto: Longmans, Green and Co., 1952.

Johnson, J. E., and P. B. Lucas. *Glorious Summer: The Story of the Battle of Britain.* London: Stanley Paul, 1990.

Kostenuk S., and J. Griffin. *RCAF Squadron Histories and Aircraft, 1924-1968.* Toronto: Samuel Stevens Hakkert and Co., 1977.

Leuthner, Stuart, and Oliver Jensen. *High Honor: Recollections by Men and Women of World War II Aviation.* Washington: Smithsonian Institution, 1989.

Mayhill, Ron. *Bombs on Target.* Sparkford: Patrick Stephens Ltd., 1991.

McAlister, Alex. *Sky-High!* Toronto: The Ryerson Press, 1944.

McGee, Eugene. *History of Port Albert.* No publisher listed, 1991.

McIntosh, Dave. *Terror in the Starboard Seat.* Don Mills: General Publishing, 1980.

Milberry, Larry, and Hugh A. Halliday. *The Royal Canadian Air Force at War, 1939-1945.* Toronto: Canav Books, 1990.

Molson, K. M., and H. H. Taylor. *Canadian Aircraft Since 1909.* Stittsville, Ontario: Canada's Wings, 1982.

Morgan, Len. *The AT-6 Harvard.* New York: Arco Publishing Co., 1965.

Peden, Murray. *A Thousand Shall Fall.* Toronto: Stoddart Publishing Co., 1988.

Pickersgill, J. W. *The Mackenzie King Record: Volume 1 1939/44.* Toronto: University of Toronto Press, 1960.

The Record. Portage la Prairie: Portage Air Observer School, 1945.

Rivaz, Richard. *Tail Gunner Takes Over.* London: Jarrolds, 1945.

Roberts, Leslie. *Canada's War in the Air.* Montreal: Alvah M. Beatty
Publications, 1943.

———. *There Shall Be Wings: A History of the Royal Canadian Air
Force.* Toronto: Clarke Irwin & Co., 1959.

Slessor, Sir John. *The Central Blue: Recollections and Reflections.*
London: Cassell and Company, 1956.

Stacey, C. P. *A Very Double Life: The Private World of Mackenzie King.*
Toronto: Macmillan of Canada, 1976.

Terraine, John. *The Right of the Line.* London: Hodder and Stoughton,
1985.

Townsend, Peter. *Duel of Eagles.* New York: Simon & Schuster, 1970.

Tredrey, F. D. *Pioneer Pilot.* London: Peter Davies, 1976.

Velleman, Alexander. *The RCAF as Seen From the Ground (A
Worm's-Eye View).* Book I: *The War Years, 1939-1945.* Stittsville,
Ontario: Canada's Wings, 1986.

Ward, Max. *The Max Ward Story.* Toronto: McClelland & Stewart,
1991.

We Serve that Men May Fly. Hamilton, Ontario: RCAF (WD)
Association, 1973.

Wynne, H. Hugh. *The Motion Picture Stunt Pilots.* Missoula, Montana:
Pictorial Histories Publishing Company, 1987.

Acknowledgements

This book could not have been written without the generous co-operation of many former students, instructors, and others involved in the British Commonwealth Air Training Plan. Their memories of that "grandiose enterprise" constitute the heart of the story; the author is grateful to:

Ernie Allen, Beckenham, England; Gerald Allester, Duncan, B.C; Tom and Marjory Anderson, Guelph, Ontario; Denis Andrews, Henley-on-Thames, England; Arthur Angus, Kitchener, Ontario; Ron Anstey, Paraparaumu, New Zealand; Chuck Appleton, Burnaby, B.C.; Lois Argue, Edmonton, Alberta.

Jack Baker, Etobicoke, Ontario; Maurice Baribeau, Pointe-Claire, Quebec; Harry Barker, Stamford, England; Jack Barnes, Scarborough, Ontario; Glenn Bassett, Calgary, Alberta; Eric Bateson, York, England; Rod Bays, Dartmouth, Nova Scotia, Marcel Beauchamp, St. Lambert, Quebec; Paul Benson, Toronto, Ontario; Bruce Betcher, Grand Forks, North Dakota; Malcolm Beverly, Willowdale, Ontario; Arthur A. Bishop, Calgary, Alberta; Arthur C. W. A. Bishop, Toronto, Ontario; R. Bradley, Ancaster, Ontario; Robert Brady, Town of Mount Royal, Quebec; Philip Bridge, Woking, England; W. Bruce Brittain, Nepean, Ontario; Gerald K. Brown, Ridgefield, Connecticut; Roy Bue, Waterloo, Ontario; Paul Burden, Fredericton, New Brunswick; John Burke, Sault Ste Marie, Ontario; William E. Burrell, Brampton, Ontario.

Allan Caine, Scarborough, Ontario; Ron A. Cassels, Gimli, Manitoba; Robert W. Caton, Swanley, England; Don Charlwood, Templestowe, Australia; Carman Harvey Chase, Willowdale, Ontario; H. D. ("Cherry") Cherrington, London, Ontario; Fred J. Chittenden, Hepworth, Ontario; William M. Clarke, London, England; A. C. Pitt Clayton, Christina Lake, B.C.; John Clinton, Mississauga, Ontario; William J. Cody, Ottawa, Ontario; Ernie Collins, Christchurch, New Zealand; Don Cooper, Willowdale, Ontario; Ewart M. Cooper, Calgary, Alberta; Roger Coulombe, Lachine, Quebec; Bill Coxhill, Milton Keynes, England; Jack Crump, Oxford, England; Gordon Roy Cudworth, Derby, England.

Don Daikens, Courtenay, B.C.; Wilfrid E. Danby, Weston, Ontario; Ralph H. Dargue, New Milton, England; Herb Davidson, Goderich, Ontario; John R. Denton, St. Catharines, Ontario; Robert J. Dixon, Boerne, Texas; Len J. Dunn, Warlingham, England; Cecil Durnin, Winnipeg, Manitoba.

Clyde B. East, Agoura, California; Arthur Eddon, Hamilton, Ontario; James F. ("Stocky") Edwards, Comox, B.C.; Jim Emmerson, Brampton, Ontario.

Neil Fletcher, Westbank, B.C.; Joe Foley, Ajax, Ontario; J. Allison Forbes, Edmonton, Alberta; C. ("Chuck") Freeman, Nanaimo, B.C.; Charles and Claire Fryer, Courtenay, B.C.; Kenneth R. Fulton, Thornhill, Ontario; Jerry Fultz, Pleasantville, Nova Scotia.

William V. Galer, Aurora, Ontario; Fraser E. H. Gardner, Agincourt, Ontario; Jake Gaudaur, Burlington, Ontario; S. W. Gerard, Richmond, B.C.; Alan Gibson, Nelson, New Zealand; Joseph W. Gibson, Don Mills, Ontario; David Goldberg, Hamilton, Ontario; P. W. Goody, London, England; Ralph W. Green, Regina, Saskatchewan; Keith E. Greenwood, Etobicoke, Ontario; Steve Gruber, Dundas, Ontario.

Arthur Hailey, Nassau, Bahamas; Ron J. Hall, Romford, England; Herb Hallatt, Campbellville, Ontario; Derek Hamilton, Goole, England; Frank Hamilton, Mazenod, Saskatchewan; Harry E. Hare, Oakville, Ontario; William S. Harker, Lethbridge, Alberta; Douglas A. Harrington, Hamilton, Ontario; Joe Hartshorn, Sarasota, Florida; W. Grant Harvey, Ottawa, Ontario; Alan F. Helmsley, Prescott, Ontario; Howard Hewer, Toronto, Ontario; John Hindmarsh, Goderich, Ontario; Harry W. Holland, Mississauga, Ontario; Jeremy Howard-Williams, Southampton, England; Lloyd B. ("Bud") Hudson, Jasper, Georgia; A. Chester Hull, Carrying Place, Ontario; Ron Hunt, Langley, B.C; Bill Hutchins, Etobicoke, Ontario; Ted Hutton, Lantzville, B.C.

Bill Irvin, Lake Placid, New York.

Alan John, Auckland, New Zealand; Edward Johnston, Barrie, Ontario; Robert A. Joss, Montreal West, Quebec.

Jim Kelly, Willowdale, Ontario; C. Roy Keys, Cannington, Ontario; Jim Kirk, Denman Island, B.C.; Hubert "Nick" Knilans, New Auburn, Wisconsin.

Ray Lackey, Currumbin, Australia; Terry Lakin, Ferndon, England; Danny Lambros, Kingston, Ontario; Reg Lane, Victoria, B.C.; Paul A. Laskey, London, Ontario; Hulbert A. Lee, Stittsville, Ontario; Wally Loucks, Etobicoke, Ontario; James Cameron Lovelace, Sydney, Nova Scotia; J. A. B. ("Jock") Lovell, Newmarket, Ontario.

Don ("Red") Macfie, Dunchurch, Ontario; John Macfie, Parry Sound, Ontario; S. J. ("Jacko") Madill, Ngaruawahia, New Zealand; Murray Marshall, Ancaster, Ontario; R. G. Marshall, Fergus, Ontario; William A. Martin, Kingston, Ontario; Ron Mayhill, Auckland, New Zealand; R. J. McBey, North York, Ontario; Joseph C. McCarthy, Virginia Beach, Virginia; Joseph P. McCarthy,

Willowdale, Ontario; C. W. McColpin, Novato, California; Ken McDonald, Willowdale, Ontario; Ray McFadden, Markdale, Ontario; James E. McInerney, Islington, Ontario; Dave McIntosh, Ottawa, Ontario; Jack McIntosh, Calgary, Alberta; James A. McPhee, Richmond Hill, Ontario; John H. McQuiston, Don Mills, Ontario; Don McTaggart, Bobcaygeon, Ontario; Jeff Mellon, Richmond Hill, Ontario; Jack Merryfield, Bragg Creek, Alberta; Walter H. Miller, London, Ontario; Dennis Miller-Williams, Halesworth, England; Albert ("Muff") Mills, Cambridge, Ontario; Ronald Monkman, Mississauga, Ontario; Len Morgan, Palm Harbor, Florida; Alan Morris, Leeds, England; Jim Morton, Burlington, Ontario; Jim Mossman, Scarborough, Ontario; Ray Mountford, Scarborough, Ontario; William B. Mouré, Calgary, Alberta; Frank Murphy, Nanaimo, B.C.

William S. Neighbors, San Antonio, Texas; Wayne Nicholls, Auckland, New Zealand; Vic Nielsen, North Bay, Ontario; Jim Northrup, Surrey, B.C.

Charles Onley, Scarborough, Ontario.

Murray Peden, Winnipeg, Manitoba; Bill Peppler, Ottawa, Ontario; Glen Phillips, Nanaimo, B.C.; Frank Phripp, St. Clements, Ontario; Wilbur ("Wib") Pierce, Mississauga, Ontario; Gerald Powell, Borehamwood, England; Keith Prior, Invercargill, New Zealand; Stephen J. Puskas, Waterdown, Ontario.

J. Alan Ramsay, London, Ontario; Charles Rawcliffe, Edinburgh, Scotland; Hugh Redfern, Huntingdon, England; Ernie Reed, Brampton, Ontario; Richard J. Reid, Mississauga, Ontario; Mac Reilley, Port Coquitlam, B.C.; Lucide Rioux, Fredericton, New Brunswick; Don Ripley, Oakville, Ontario; Harry Ritchie, Brampton, Ontario; Jack Ritchie, Oakville, Ontario; Earl William Roberts, Fenelon Falls, Ontario; Kenneth G. Roberts, Ottawa, Ontario; Thomas A. Robinson, Ottawa, Ontario; Cecil A.

Robson, Saskatoon, Saskatchewan; Jeffrey N. Rounce, Great Walsingham, England; Jim Ruddell, Osgoode, Ontario; Harry Russell, Auckland, New Zealand.

John J. Saqui, Ladysmith, B.C.; Raymond Savage, Calgary, Alberta; Rayne D. ("Joe") Schultz, Ottawa, Ontario; Norman Shrive, Burlington, Ontario; John H. Simpson, Ottawa, Ontario; Don Sinclair, Bury St. Edmunds, England; J. R. Sipple, St. Petersburg Beach, Florida; James M. Smythe, Newmarket, Ontario; Gordon C. Southcott, Mississauga, Ontario; Clifford J. Stead, Kanata, Ontario; George Stewart, Hamilton, Ontario; Eric Stofer, Victoria, B.C.; William H. Swetman, Tottenham, Ontario.

Alfred Tait, Fredericton, New Brunswick; Richard Taylor, Victoria, B.C.; Norman Thacker, Oakville, Ontario; Lucien Thomas, Scottsdale, Arizona; Walter F. Thomson, Fort Assiniboine, Alberta; Alison Tucker, London, Ontario; John C. Turnbull, Toronto, Ontario; Allan Turton, Kettleby, Ontario.

Douglas Wadham, Consett, England; Bill Wagner, Whittier, California; D. R. ("Danny") Walker, London, Ontario; A. B. ("Tet") Walston, Comox, B.C.; Douglas ("Duke") Warren, Comox, B.C.; John Watt, Ohope, New Zealand; George James Webb, Santa Cruz, California; Phil G. Weedon, Westerose, Alberta; William Weighton, Victoria, B.C.; William S. White, Nelson, New Zealand; Norman Whitley, Comox, B.C.; Vernon A. Williams, Oakville, Ontario; Ross C. Wiens, Ville d'Anjou, Quebec; Joffre Woolfenden, London, Ontario; F. Peter Wrath, London, Ontario; John A. Wullum, Surrey, B.C.

Special thanks go to Doug Harrington and Norman Shrive for their invaluable assistance in the development of the section on navigation training; to J. Alan Ramsay, who was equally helpful on gunnery and bombing; and to Grant Harvey, Chester Hull, Ken

McDonald, John Simpson, and Tet Walston, who were notably generous with their time and their memories of a remarkable period in Canadian history. In addition, the author conveys his appreciation of the assistance provided by Owen Cooke and several others at the Directorate of History, Department of National Defence, Tim Dubé, Sarah Montgomery, and Pat Robertson of the National Archives of Canada, Ralph Leonardo of the National Aviation Museum, Earl Hewison of the RCAF Memorial Museum, Trenton, Harry Hayward of the Commonwealth Air Training Plan Museum, Brandon, Manitoba, Jack Evans and Rob Schweyer of the Canadian Warplane Heritage Museum, Mount Hope, Ontario, Major K. W. Farrell of Canadian Forces Photographic Unit, Ottawa, as well as the staff at the Western Canada Aviation Museum, Winnipeg, Manitoba, and at the Comox Air Force Museum, Lazo, B.C., whose generous assistance made the author's research a genuine pleasure.

Thanks also to Jeanne Muldoon of London, Ontario, who made available her extensive files on the training fields in the London area and provided introductions to scores of former students and instructors. Rod Priddle of Devizes, England, generously furnished many excerpts from letters written by his late uncle, Phillip Del Rosso, who trained with the BCATP. Joe Clark of King City, Ontario, was a fund of information about his late father and the machinations that led to the BCATP's starring role in Hollywood. John and Margaret Inglis of Mossbank, Saskatchewan, went to a great deal of trouble to find out about the impact of No. 2 B&GS on the local community during the war years. Stuart Ng of the School of Cinema-Television, University of Southern California, was most helpful in making Warner Brothers archival material available to the author. To all these individuals – and to the Ontario Arts Council, Lucinda Vardey, and Doug Gibson, Alex Schultz, and Lynn Schellenberg of McClelland & Stewart – sincere thanks.

Indexes

THE BRITISH COMMONWEALTH AIR TRAINING PLAN
AND ITS UNITS

GENERAL INDEX